T0314138

Ultrasound Technology for Clinical Practitioners

Ultrasound Technology for Clinical Practitioners

Crispian Oates
Newcastle University, UK

WILEY

Registered Office
John Wiley & Sons, Inc., 111 River Street, Hoboken, NJ 07030, USA
John Wiley & Sons Ltd, The Atrium, Southern Gate, Chichester, West Sussex, PO19 8SQ, UK

For details of our global editorial offices, customer services, and more information about Wiley products visit us at www.wiley.com.

Library of Congress Cataloging-in-Publication Data Applied for:
ISBN 9781119690825

Cover Image: Courtesy of the Author
Cover Design: Wiley

Set in 10.5/13pt STIXTwoText by Straive, Pondicherry, India

C9781119891550_191222
Printed and bound by CPI Group (UK) Ltd, Croydon CR0 4YY

This book is dedicated to Tony Whittingham who has been my guide and mentor in all things ultrasonic and to Marion my constant companion and encourager.

Contents

Acknowledgements

In writing this book, I have been helped by many friends and colleagues. I am very grateful for all their contributions and help, and indebted to them for their advice. It began with interaction with my students over many years and a set of honed lecture notes. More specifically, I have been particularly helped by Gareth Bolton at the University of Cumbria who very kindly let me use his facilities and scanners to get a significant number of the clinical images demonstrating machine settings. He also gave me valuable feedback, reading drafts and looking at the material from the student-teacher point of view. Stephen Klarich and Jamie Wild gave me invaluable advice on the practicalities of using elastography and Chris Eggett advised me in relation to echocardiography. Barry Ward advised on quality assurance. I am also thankful to Kathia Fiaschi, Carmel Moran, and Heather Venables. Those mentioned and others kindly provided images as indicated throughout the book. Putting together a book like this takes a considerable time and Covid lockdown certainly helped in providing that time. So too did my family and particularly my wife Marion. She has been a great support and encourager and I am deeply thankful to her. However, having acknowledged those who have helped and contributed in various ways, the work is mine as are any errors and mistakes that remain within it.

Crispian Oates
April 2020

List of Abbreviations

ADC	analogue to digital converter
AI	artificial intelligence
A-mode	amplitude mode
ARFI	acoustic radiation force impulse
B-flow	B-mode flow
B-mode	brightness mode
CA	contrast agent
CAD	computer aided diagnosis
CCA	common carotid artery
CDU	colour Doppler ultrasound (CFM)
CEUS	contrast enhance ultrasound
CFM	colour flow mapping (CDU)
CLA	curvilinear array
CMUT	capacitative micromachined ultrasound transducer
CPU	central processing unit
CT	computerised tomography
CUTE	computed ultrasound tomography in echo mode
CW	continuous wave
DGC	depth gain control (TGC)
ECG	electrocardiogram
FFT	fast Fourier transform
FPS	frames per second
FR	frame rate
FWHM	full width at half maximum (beamwidth)
GPU	graphics processing unit
ICA	internal carotid artery
I_{SATA}	spatial average temporal average intensity
ISB	intrinsic spectral broadening
I_{SPPA}	spatial peak, peak average intensity

I_{SPTA}	spatial peak temporal average intensity
I_{SPTP}	spatial peak temporal peak intensity
IUCD	intra-uterine contraceptive device
LA	linear array
MI	mechanical index
M-mode	motion mode
MRI	magnetic resonance imaging
PA	phased array
PD	power Doppler
PI	pulsatility index
PRF	pulse repetition frequency
PSV	peak systolic velocity
pSWE	point shear wave elastography
PVDF	polyvinylidene flouride
PWD	pulse wave Doppler
PZT	lead zirconate titanate
QA	quality assurance
RBC	red blood cell
RF	radio frequency
RI	resistance index
ROI	region of interest
RSI	repetitive strain injury
Rx	receive (signal)
SA	synthetic aperture
SCA	subclavian artery
SE	strain elastography
SNR	signal to noise ratio
SoS	speed of sound
SR	strain rate
SRT	systolic rise time
SSI	supersonic shear (wave) imaging
STE	speckle tracking echocardiography
SV	sample volume
SWE	shear wave elastography
TDI	tissue Doppler imaging
TGC	time gain control (DGC)
TI	thermal index
TIB	thermal index for bone in view
TIC	thermal index for superficial bone in view

TIS	thermal index for soft tissue
Tx	transmit (signal)
UFCD	ultrafast colour Doppler
UFUS	ultrafast ultrasound
VFI	vector flow imaging
WRRSI	work related repetitive strain injury

Introduction

This book covers the essential physics and technology of diagnostic ultrasound needed by someone practicing ultrasound in the clinical setting, with ultrasound as a primary or significant component of their job. For simplicity, the term 'sonographer' has been used throughout for this person but ultrasound is used by a wide range of personnel in clinical practice including doctors, echocardiographers, vascular scientists, midwives, nurse practitioners, and physiotherapists. The book is designed to be accessible to all of these practitioners. Each chapter is liberally illustrated with easily reproducible drawings and clinical images to demonstrate the point being made. The use of equations has been kept to a minimum. Where used, equations are useful in showing the relationship between one factor and another and where changing one thing can have clinical or safety implications. The term 'scanner' refers to the ultrasound machine.

Over the years, ultrasound machines have become more user friendly and the machine performs many functions without the user being aware of what is being changed, for example the use of presets for particular patient examinations. It is important to have an understanding of what your equipment, being applied to a patient, is doing. By knowing more about the technology behind the scanner, you will increase in confidence in handling your scanner and be able to be assured that you are obtaining the optimum diagnostic information and are operating in a safe manner for yourself and the patient. The patient may also be reassured that they are being treated by a competent practitioner who knows their equipment and what they are doing.

The emphasis throughout will be on what the user needs to know in order to drive the ultrasound scanner correctly and effectively in order to obtain the best images. We will also look at the technical factors that must be taken into account when interpreting the images to make clinical judgments. In a number of places, it will be necessary or useful to explain a point in greater detail or add additional but less essential information. These will be indicated by using green shaded boxes. Where a point of specific relevance to daily practice is made, green text is used. Key terms are highlighted in bold type.

Ultrasound uses sound waves of a higher pitch than the audible range to form images from within the body. The ultrasound is produced by a **transducer probe** that is typically placed on the skin, after it has had a liquid gel applied to it. Short pulses

Ultrasound Technology for Clinical Practitioners, First Edition. Crispian Oates.
© 2023 John Wiley & Sons Ltd. Published 2023 by John Wiley & Sons Ltd.

of ultrasound are transmitted into the body and are reflected, forming echoes that in turn are picked up by the ultrasound probe. The received signal is then processed to form the image we see. This method of forming images is called the **pulse-echo technique** and is similar to that used by sonar on boats or by radar.

In Figure 0.1, we see three typical ultrasound greyscale images. Looking at the three images, the first thing to notice is that there are two basic image formats. A rectilinear format or **linear scan** (a) and a sector shaped format or **sector scan** (b,c). Both formats show a cross-sectional slice through the body as though the body had been cut open, going deep from the transducer probe at the skin surface, and we are looking down onto the cut surface (Figure 0.2).

As shown, the transducer/skin surface is at the top of the linear image and at the narrower point of the two sector images. The orientation of the image may be chosen by the user. The three images are produced by different ultrasound probes and

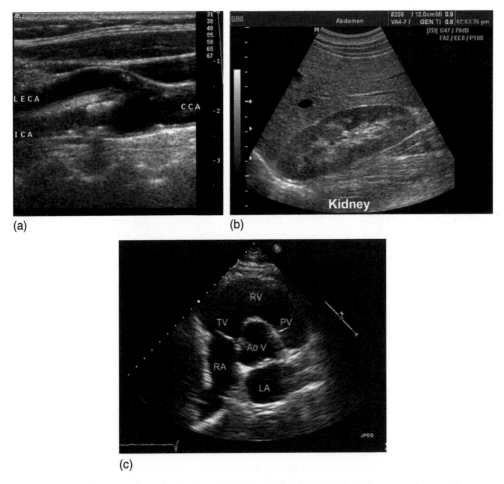

(a)

(b)

(c)

FIGURE 0.1 Three typical greyscale or B-mode images from (a) a linear array, (b) a curvilinear array and (c) a phased array. Image (b) shows the greyscale used down the left-hand side.

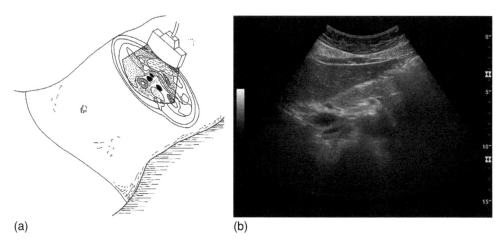

(a)   (b)

FIGURE 0.2   The image plane within the body (a) and the ultrasound image seen (b).

are useful for different clinical examinations. In the first image (a), the linear scan is produced by a **linear array probe** and is useful for looking at small parts including musculoskeletal examinations and vascular work. The second image (b) is produced by a **curvilinear probe** and is useful for situations where an extended target is being viewed, for example the abdomen or a foetus. The third image (c) is produced by a **phased array probe**. This probe has a small footprint on the skin but is able to show an extended field of view in the body, for example viewing the heart from between the ribs.

At its most basic, the process of image formation can be thought of as using a narrow beam of ultrasound to sweep through the tissue across the **image plane** or **scan plane,** like we might sweep a torch beam across a dark room, to build up a picture of what is there. The image is therefore built up from a series of lines transmitted out from the transducer and going deep into the body, laid side-by-side to form the image. Each line uses the pulse-echo principle to receive echoes from along that line. In the case of the linear scan, it is as though the beam was swept in a straight line along the surface of the skin across the width of the probe. In the case of the phased array, it is as though the probe was rocked at one point on the skin to sweep the beam in a sector.

If we know how fast the pulse travels through the tissue, we can time the echoes coming back and so calibrate the depth of the echo targets in centimetres away from the probe. The marks down the side of each image indicate depth from the transducer probe.

Looking at the images, details of structures producing echoes are shown as bright and dark marks matching a greyscale as seen on the left side of Figure 0.1b. This type of image is known as a **B-mode** or **greyscale image**. Some parts of the image are clearer and more obvious than other parts. In order to interpret these images, the scanner must firstly be set up to produce the optimum quality image and secondly, the person interpreting needs to understand what is really being imaged and what is **artefact** or misrepresented in the image. As for all imaging modalities, the sonographer needs anatomical knowledge and an understanding of how the images are formed and what limitations the technique has in order to correctly interpret the images.

Whilst modern machines have automated many of the processes involved in ultrasound imaging, there remain a large number of variables under user control that the sonographer must manage effectively to produce optimal images and maximise the diagnostic potential of ultrasound.

We begin our look at ultrasound technology by considering what ultrasound is and how it interacts with tissue. We then move on to look at the production of a B-mode or greyscale image and its interpretation before going on to consider Doppler ultrasound. We then look at making measurements, safety of ultrasound, and quality assurance before moving on to advanced topics and the latest developments with ultra-fast techniques. Finally, we look at elastography. Three appendices cover a check list for performing a scan that covers 'knobology', the basic manipulation of equations, and a detailed look at the ultrasound beam.

CHAPTER 1

The Basic Physics of Ultrasound

SOUND WAVES

A sound wave is a fluctuating variation in pressure within a medium such as air, water, or solid material. Our ears are sensitive to such pressure changes in air, and we hear sounds all the time. The faster the changes in pressure take place, the higher the pitch or **frequency** of the sound we hear. Frequency is measured in hertz (Hz) and, for a young person, their hearing goes from 20 Hz to 20 kHz. Middle C on a piano is 261 Hz. A sound above 20 kHz is called **ultrasound** ('beyond sound'), in other words, we cannot hear it. Dogs can hear sounds of higher frequency than we can, and bats use ultrasound, up to 200 kHz, for echo location in the dark. The ultrasound we use for medical imaging is in the range of megahertz (MHz), far above anything we can hear. As we will see, the reason for going to such high frequencies is that we can then make narrow beams of ultrasound that we can point in a particular direction and which can produce high-resolution images showing fine details.

We can think of a medium like air or water as a collection of molecules with mass connected by springs that represent the forces between the molecules (Figure 1.1). If you push on one molecule, it will move closer to the adjacent molecule and exert a force so that it too begins to move. Adjacent molecules having been squashed together will then repel one another and recover to their resting position. They will keep moving beyond their resting position due to their momentum. The force holding them together then becomes an attractive force that pulls the molecules back towards

Ultrasound Technology for Clinical Practitioners, First Edition. Crispian Oates.

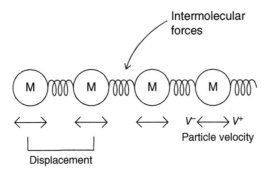

FIGURE 1.1   Ball and spring model of a sound wave travelling in some medium.

their resting position again as they are then stretched apart. That is, each molecule will move forwards then backwards in the direction of the applied force. This oscillating molecular motion within a material is the basis of a sound wave.

DEFINITION

A **sound wave** is a longitudinal pressure wave travelling at the **speed of sound** through a medium (e.g. air, water, and soft tissue).

The pressure is the force exerted from the excess density (number of molecules per unit volume) of molecules above or below the average density of the medium as the molecules get alternately squashed and pulled apart from one another. In other words, it is the excess pressure about the mean resting pressure.

A sound wave may be generated by a piston moving forwards and backwards. This is what a normal loudspeaker does (Figure 1.2). It has a diaphragm that moves forwards and backwards driven by an electric signal. This will alternately push, then move away, from the molecules in front of it. These molecules will then alternately push on to those in front of them and then pull on them, and so on, as described

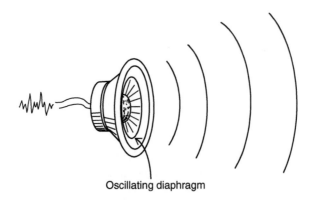

FIGURE 1.2   A loudspeaker driven by an electric signal acts as a piston on the air in front of it.

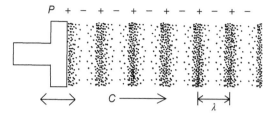

FIGURE 1.3 Oscillating piston producing a longitudinal pressure wave, a sound wave consisting of compression (+) followed by rarefaction (−). c is the speed of sound, and λ is the wavelength.

above. This will create a series of compressions and rarefactions in the molecules of the medium that move away from the source of movement (the piston), i.e. a sound wave as shown in Figure 1.3.

The speed at which they move away from the source is called the **speed of sound** (c). The speed of sound can be considered constant for each medium, so the speed of sound in air is 330 m·s⁻¹, and if a plane goes faster than this it breaks the sound barrier. The speed of sound in water varies with temperature and at 20 °C is 1480 m·s⁻¹. The average speed of sound in soft tissue is 1540 m·s⁻¹, and, as will be seen, this is a key number for ultrasound imaging.

For ultrasound imaging, the ultrasound transducer acts exactly like the loud-speaker pushing and pulling the molecules of the medium in front of it.

NOTE

The individual molecules oscillate backwards and forwards about a mean position, but the pressure disturbance (p) propagates forward at the speed of sound (c). It is the moving disturbance that is the sound wave.

A typical sine wave plot of a sound wave is seen if we plot the change in pressure (p) at a given point against time, or if we plot the change in pressure versus distance away from the piston.

Looking at one position in space versus time (t), we see the pressure increasing and decreasing as the sound wave passes by (Figure 1.4a).

The **frequency** (f) is the number of cycles (peaks) passing a given point in one second.

The **period** (T) is the time taken to complete 1 cycle. The relationship between period and frequency is

$$f = \frac{1}{T}$$

Looking at one instant of time versus distance (x) away from the sound source (Figure 1.4b), we see the pressure increases and decreases as we move away from the transducer.

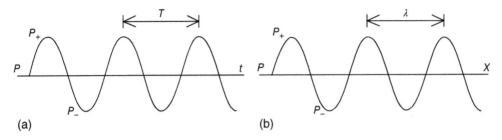

(a)   (b)

FIGURE 1.4 Illustration of the change in pressure with time at one point in space (a) and the change in pressure with distance x from the sound source (b).

The **wavelength** (λ) of the sound is then defined as the distance in space between two successive peaks on the wave.

THE SPEED OF SOUND EQUATION

The relationship between speed of sound (c), frequency (f), and wavelength (λ) is

$$c = f.\lambda$$

Relationship Between Pressure, Particle Velocity and Particle Displacement

We can also plot the change in particle velocity (v) versus time and the particle displacement (s), from its resting position, versus time, to give similar graphs (Figure 1.5).

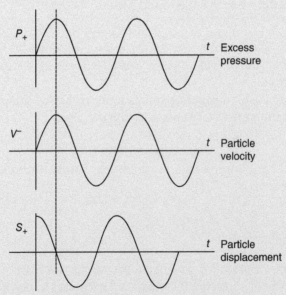

FIGURE 1.5 The sine waves produced by plotting the change in pressure, particle velocity, and displacement associated with a sound wave.

Comparing these waves, we see that the pressure is greatest when the particle velocity is greatest and the particle displacement is greatest when the rising pressure passes through its mean zero level.

When we simply talk about a 'sound wave,' we usually mean the (excess) **pressure wave** – also known as **acoustic pressure**. It is acoustic pressure that our ultrasound transducers are sensitive to and detect.

NOTE

Do not confuse particle velocity with the speed of sound. Particle velocity is movement at a molecular level, whereas the speed of sound is the speed at which the sound wave propagates through the medium.

A **transducer** is anything that converts one form of energy into another form. The loudspeaker and the ultrasound transducer both convert electrical energy into sound energy and so are transducers.

NOTES

- As a sound wave propagates its frequency remains constant at the same frequency as the transducer 'piston' oscillates.
- For a given medium, the speed of sound is constant (it may vary with temperature). This means that if the transmitted frequency increases, the wavelength must get shorter to balance the speed of the sound equation.
- **Key Concept**: 'High frequencies give short wavelengths'.

DESCRIBING WAVES

The **amplitude of a wave** is the difference between the peak value and the mean zero value. We can also define the **peak-to-peak amplitude** A_+ to A_- as shown in Figure 1.6.

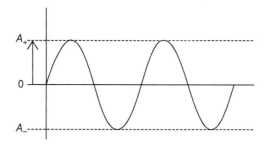

FIGURE 1.6   Definition of wave amplitude A.

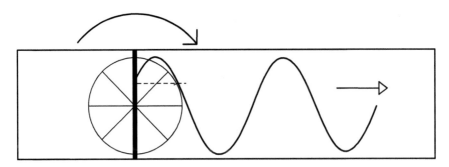

FIGURE 1.7   Illustration of a spinning bicycle wheel over a moving roll of paper.

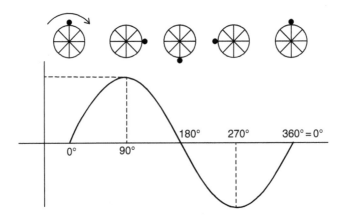

FIGURE 1.8   Definition of the phase angle of a sound wave.

The **phase of a wave** is a point along the course of one period of the wave expressed as an angle.

Think of a spinning bicycle wheel mounted above a moving roll of paper as shown in Figure 1.7. On the paper, we mark the vertical distance of the valve away from the hub at each moment as the wheel spins round. What results is a sine wave drawn on the paper. By the time, the wheel has gone round once, you would have drawn one period of the sine wave and the valve would have travelled round 360°. So, by measuring the angle of the valve as it goes round, we can mark the phase angle along the sine wave as shown in Figure 1.8. One cycle is equal to 360° (hence, frequency is equal to 'cycles' per second).

This gives us a very useful way to compare two sine waves. If one wave has a phase angle of 45° at the same time another sine wave has a phase angle of 0°, we know where the peak of one wave is compared to the other (Figure 1.9). If we know the frequency, amplitude, and phase of a wave, we know everything about it. The wavelength will depend on the speed of sound of the medium the sound wave travels through.

A special case of sound wave transmission is seen if we look at a single frequency sound wave propagating in one direction through a uniform medium from an infinitely large sound source. This type of wave is called a **plane wave**. The **wavefronts**

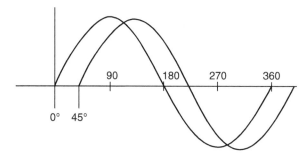

FIGURE 1.9   Two sine waves with a phase difference of 45°.

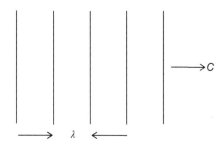

FIGURE 1.10   A plane wave showing the wavefronts parallel to one another.

(wave peaks) are then parallel to one another (Figure 1.10). Sea waves coming on to a straight beach closely approximate this situation.

Plane Waves

If we compare the pressure wave with the particle displacement and particle velocity waves for a plane wave, they all have the same frequency. Looking at Figure 1.5 and comparing the phase difference between them, we find that the excess pressure and particle velocity waves are in phase with each other (i.e. peaks occur at the same time), whilst the particle displacement wave is 90° ahead of the pressure wave. That is, its peak occurs a quarter of a cycle before that of the pressure and particle velocity waves. Their relative amplitudes will depend on the medium they are travelling through.

ENERGY IN A SOUND WAVE

In order to move the sound source piston backwards and forwards, work must be done and this requires energy. The disturbance created in the molecules can then make molecules some distance away from the piston move, so the sound wave must be transporting energy through the medium. The average pressure in a sound wave is zero as the amplitude oscillates equally above and below the mean. But the amplitude squared

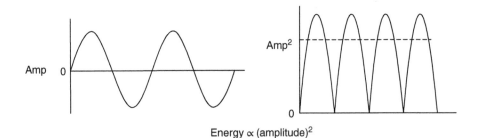

Energy ∝ (amplitude)²

FIGURE 1.11   The relationship between the amplitude of a sound wave and the energy carried by the sound wave.

always has a positive value (think of $(-1) \times (-1) = +1$) (Figure 1.11). The **energy** carried by a sound wave is proportional to the pressure amplitude squared (amp)².

So, for the acoustic pressure p, energy $E \propto p^2$.

The energy carried by a sound wave is important when considering the safety of ultrasound exposure in the body. The ultrasound is depositing energy in tissue, and it can cause damage if too much energy is deposited in one place.

We will consider the energy and safety of ultrasound further in Chapter 11.

ULTRASOUND PULSES

A pulse is a very short burst of sound. Most of the imaging we do uses very short pulses of sound, so we can tell where the echoes are coming from at any one time.

If we consider a pulse in time, it will have a shape and duration. A typical imaging pulse might have a large amplitude at first then die away (Figure 1.12).

- Peak amplitude – typically, 1 atmosphere $= 10^5$ Pa
- Particle velocity – typically, peak amplitude value ~ 10 cm s⁻¹
- Particle displacement – typically, peak amplitude value ~ 20 nm (1 nm $= 10^{-9}$ m)
- Pulse duration – three—four cycles for B-mode ultrasound at 5 MHz
- Pulse duration – 0.2 μs × 4 = 0.8 μs

We can also consider the shape of the pulse in space at one instant in time. Note the reversal of wave shape – peak amplitude comes first in time and in space (Figure 1.13).

- Pulse length in tissue – three—four cycles
- For B-mode ultrasound at 5 MHz in soft tissue
  $c = 1540\,\text{m·s}^{-1}$ $\lambda = 0.3\,\text{mm} \times 4$ to give a pulse length of 1.2 mm

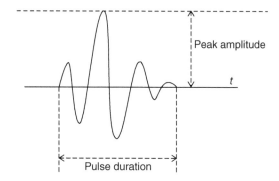

FIGURE 1.12   Typical pulse shape of a short pulse of ultrasound.

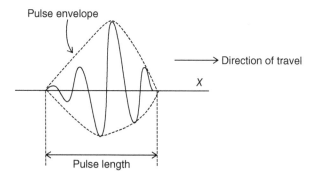

FIGURE 1.13   The pulse envelope of an ultrasound pulse in space.

We can characterise the shape of the pulse by drawing a line through successive peaks to show the **pulse envelope,** as shown by the dashed line.

ENERGY SPECTRUM OF A PULSE

DEFINITION

A **spectrum** is a plot of energy against frequency.

- For example, in a light spectrum, as seen in a rainbow, red has a lower frequency and blue has a higher frequency. The energy is seen in the brightness of the colours.

The rate at which energy is produced or transmitted is called **power,** which is equal to energy per unit time ('**rate**' means 'per unit time').

For a sound wave, the **power spectrum** is a plot of the power in the sound wave versus the frequency of the wave. Figure 1.14 shows what the spectrum of a simple sine wave of frequency f_0 looks like.

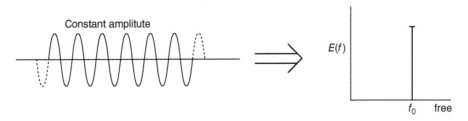

FIGURE 1.14   The single line power spectrum of a continuous sound wave of a single frequency, f_0.

It has all of its energy (E_0) at a single frequency (f_0).
For a pulse, we can ask the following questions:

- What is the frequency of a pulse?
- What does the spectrum of a single pulse look like?

BANDWIDTH

Looking at the pulse we saw in Figure 1.12 or the pulse in Figure 1.15, we see that no two cycles are the same, unlike the pure sine wave where every cycle is identical. On the other hand, the waveform for most of the pulse itself does look like a sine wave. So, what is its frequency?

The answer is that the pulse has a spectrum that centres around f_0, but extra frequencies need to be included on either side to account for the sine wave stopping at the ends of the pulse. Figure 1.16 shows the spectrum of the pulse has a **bandwidth** or range of frequencies centred about f_0. We talk about the **'full width at half maximum'**, or **FWHM**, as a measure of the bandwidth. This is the width of the spectrum, in hertz, measured at the half height of the peak around the centre frequency.

We can then look at the shape a pulse has and we can consider the energy that the pulse carries. The energy of a pulse becomes an important factor when we look at the safety of ultrasound.

When talking about the shape of a pulse over time, we talk about the **time domain** and when talking about the power spectrum of the pulse, we talk about the **frequency domain.** Figure 1.17 shows what the energy spectrum is for various pulses, i.e. the frequency content of a single pulse.

Notice how the very long pulse looks more like the sine wave than the very short pulse. Similarly, its spectral bandwidth is narrow and close to that of the sine wave frequency, f_0. The shorter the pulse, the wider the bandwidth of the spectrum, until with an infinitesimally short pulse the bandwidth is infinite and there is no associated single frequency, f_0. All frequencies are equally present. As an example of an infinitesimally short pulse, in order to study how all the frequencies respond at once in a concert hall, sound engineers use a gunshot from a starting pistol. A sharp hand clap also has a very wide range of frequencies emitted.

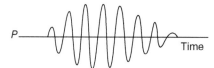

FIGURE 1.15 A 7-cycle pulse.

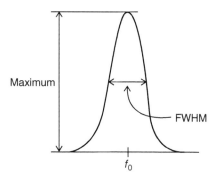

FIGURE 1.16 The full width at half maximum or FWHM bandwidth of an ultrasound pulse with centre frequency, f_0.

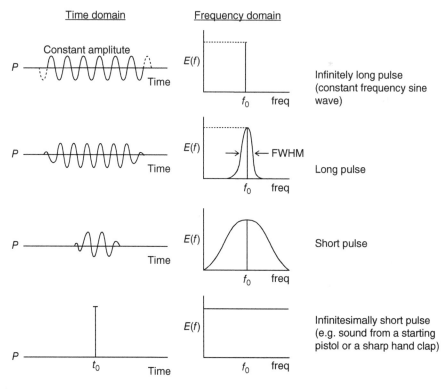

FIGURE 1.17 The relationship between the time domain picture and the frequency domain picture for various pulse shapes.

SUMMARY

There is a reciprocal relationship between the time domain and the frequency domain. An infinitely long sine wave has a single frequency whilst a single instantaneous sound has infinite bandwidth.

The importance of this relationship for medical ultrasound is that in order to have very short pulses and to give high-resolution images, we must use transducers that have a wide bandwidth so they can transmit and receive across a wide range of frequencies. For this reason, manufacturers will emphasise the fact that they have **wideband transducers**. Typical pulse shapes for B-mode and Doppler pulses are shown in Figure 1.18.

An approximation of the bandwidth of a given pulse length is given by

$$\text{Bandwidth} \sim \pm 2 \times \frac{1}{\text{Pulse length}}$$

So, a short pulse gives a wider bandwidth. For a typical diagnostic pulse of three cycles for B-mode, FWHM bandwidth, $\approx f_0$, where f_0 is the centre frequency. For example, a 5 MHz pulse has an FWHM bandwidth of ± 2.5 MHz

Quality Factor

Some texts refer to the quality factor Q of a resonator such as an ultrasound transducer (not to be confused with Q for flow volume!)

$$Q = \frac{\text{Centre frequency}, f_0}{\text{Bandwidth}}$$

So, a wideband transducer has a low Q, and a transducer with a narrow bandwidth has a high Q.

SPEED OF SOUND (C)

(Note: do not confuse the speed of sound with particle velocity!)

Returning to the ball and spring model of a material, Figure 1.19, we see that the model consists of a series of masses representing the molecules, connected together by springs of a certain stiffness representing the forces between the molecules. In

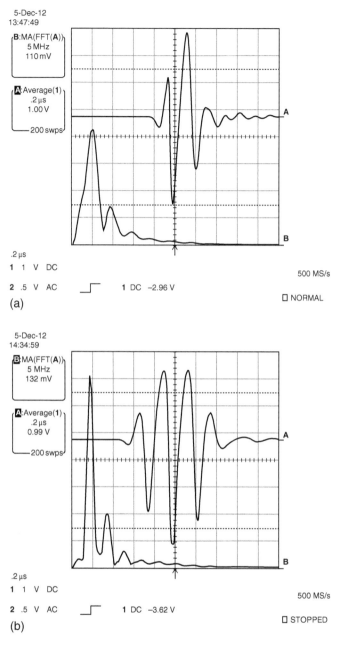

FIGURE 1.18 (a) A B-mode pulse and (b) a pulse wave Doppler pulse. The lower left corner of each image shows the spectrum of each pulse. Note that the longer pulse in (b) has a narrower bandwidth. *Source:* Courtesy: Jie Tong.

order to get a sound wave to propagate through the medium we need to get a mass moving and that moving mass must transmit its movement onto the next molecule through the spring. The lighter the mass, the easier it will be to get it moving, and the

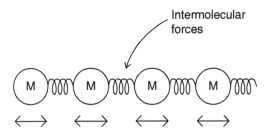

FIGURE 1.19   Ball and spring model of the molecules of a material showing the inter-molecular forces between the oscillating molecular masses.

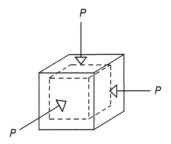

FIGURE 1.20   Definition of the bulk modulus showing the decrease in volume with an increase in applied pressure, *P*.

stiffer the spring, the faster it will transmit that movement on to the next molecule. We may, therefore, expect that light stiff materials will have a high speed of sound, and dense soft materials have a low speed of sound. For example, aluminium, a light rigid metal, has a speed of sound 5100 m·s⁻¹, whilst in air the molecules are also light but are very poorly connected, and the speed of sound is only 330 m·s⁻¹.

In terms of what we can measure, the average mass per unit volume of material is called the **density** ρ (kg·m⁻³), and the **stiffness** is a measure of the forces between the molecules

For a volume of a material, its resistance to such forces is called the **bulk modulus** *K*.

The bulk modulus relates the fractional change in volume $\Delta V/V$ to the applied pressure ΔP (Figure 1.20).

$$\text{Bulk modulus, } K\left(K\right) \qquad K = -V\frac{\Delta P}{\Delta V}\left(\text{kg}\cdot\text{m}^{-1}\cdot\text{s}^{-2}\right)$$

The minus sign occurs because the volume decreases with an increase in pressure.

Speed of sound in a material, $c = \sqrt{\dfrac{K}{\rho}}$

> **NOTES**
> - At a given temperature, the speed of sound is constant.
> - Speed of sound is independent of frequency (for most purposes in medical ultrasound).
> - Materials differ more in their stiffness than in their density. Therefore, stiffness is a better guide to predicting c than density.

Looking at Table 1.1, we see that the speeds of sound for different soft tissues are quite close together. The average speed of sound in soft tissue in the body is $1540\,\text{m·s}^{-1} \pm 5\%$. This is close enough to $1540\,\text{m·s}^{-1}$ for us to use this value in most circumstances in clinical ultrasound. The value **$1540\,\text{m·s}^{-1}$ is a key number to remember when considering how ultrasound propagates through soft tissues**. The variation in speed of sound from $1540\,\text{m·s}^{-1}$ will produce artefacts in the image (see Chapter 7). However, sometimes we need to apply a more accurate value to our measurements, for example when making measurements of the eye in ophthalmology. In Chapter 13, we will see how such variations in speed of sound may be corrected to improve image quality.

TABLE 1.1   Showing the Speed of Sound for Various Materials

Speed of sound table		
Material	**Speed of Sound (m·s⁻¹)**	
Air	330	
Water	1480	
Plastic (Perspex)	2730	
PZT transducer	3741	
Fat	1450	
Brain	1546	Soft tissue average
Liver	1550	
Kidney	1560	$1540 \pm 5\%$
Blood	1570	
Muscle	1580	
Bone (cortical)	3500	

CHARACTERISTIC ACOUSTIC IMPEDANCE, Z_0

> **QUESTION**
>
> How easily does a molecule in the medium move in response to a given change in acoustic pressure?

The answer is given in a quantity called the **characteristic acoustic impedance, Z_0**. This is a measure of how easily a molecule in the medium moves in response to the excess pressure (p) of the sound wave – assuming the sound wave is a plane wave.

Z_0 is constant for a given medium.

Characteristic Acoustic Impedance

At a molecular level, acoustic impedance $Z_0 = \dfrac{p}{v} = \dfrac{\text{Excess pressure}}{\text{Particle velocity}}$

Recall that for a plane wave, the pressure wave and the particle velocity are in phase with one another. The ratio of their amplitudes will therefore be constant at all points in the wave and equal to Z_0 as shown in Figure 1.21. This ratio is the **acoustic impedance** of the material carrying the sound wave.

Z_0 is constant for a given medium.

Think of it like the gears on a bike:

- The force exerted on the pedal ≡ pressure
- Speed of the pedals ≡ particle velocity

High gear   Large pressure and low response to pedal push
Low gear   Small pressure gives big response to pedal push

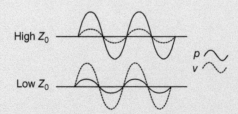

FIGURE 1.21   The relationship between pressure amplitude and the particle velocity for two materials with a large and small acoustic impedance Z_0.

> **KEY CONCEPT**
>
> Acoustic impedance Z_0 is important because **it is the difference in Z_0 between different materials that determines the size of the echo at an interface.**

Table 1.2 shows the value of Z_0 for a range of materials.

We cannot measure the excess pressure and the particle velocity of molecules in a material very easily. In terms of what we can measure, Z_0 depends on the stiffness (bulk modulus) K and density ρ of the medium and can be calculated by measuring the density and speed of sound, c.

$$Z_0 = \rho \cdot c = \rho \sqrt{\frac{K}{\rho}} = \sqrt{\rho K}$$

A dense, rigid material will have a large Z_0, whilst a less dense non-rigid material will have a low Z_0.

> **NOTE**
>
> Many texts refer to acoustic impedance as the 'resistance' to the sound wave by the medium. Resistance is usually associated with energy loss, for example, electrical resistance causing heating of a wire. However, when considering echoes, acoustic impedance is not associated with any energy loss. Understanding that acoustic impedance is the 'response of the molecules' to the sound wave is much to be preferred as it is then not confused with the energy losses that do occur as sound travels through a medium. The energy losses are due to attenuation of ultrasound which will be discussed in Chapter 2 (in electrical terms, acoustic impedance is the analogue of reactance).

TABLE 1.2   Speed of Sound, Density, and Acoustic Impedance Z_0 for Several Tissues and Materials

Material	Speed of Sound $(m \cdot s^{-1})$	Density $(kg \cdot m^{-3})$	Z_0 $(kg \cdot m^2 \cdot s^{-1} \times 10^6)$
Air	330	1.3	0.0004
Water	1480	998	1.48
Plastic (Perspex)	2730	1180	3.22
PZT transducer	4560	7750	35.34
Fat	1450	911	1.32
Liver	1550	1079	1.67
Kidney	1560	1066	1.66
Blood	1570	1057	1.66
Muscle	1580	1090	1.72
Bone (cortical)	3500	1908	6.68

ENERGY IN A SOUND WAVE

Energy (joule $J = kg\,m^2s^{-2}$)   (recall: energy $\propto p^2$, where p is acoustic pressure)

Power (watt, W or $J\cdot s^{-1}$) is the rate at which energy is transferred.

Total transmitted power is the total sound energy emitted by the transducer summed over all directions.

Recall: The sound source is a piston doing work moving backwards and forwards. It makes the molecules in front move, and energy is propagated away from the transducer as a sound wave.

Energy is not generally transmitted equally in all directions. For example, we want a narrow beam of ultrasound for imaging. We, therefore, need to know how much energy is being transferred in each direction away from the transducer. For this, we need to know the **intensity** of ultrasound.

Intensity ($W\cdot m^{-2}$ or $mW\cdot m^{-2}$ or $J\cdot s^{-1}\cdot m^{-2}$ or $kg\cdot s^{-3}$) is the power per unit area. It is the energy flowing across an imaginary surface cutting across the ultrasound beam as shown in Figure 1.22.

One way of thinking about intensity is to think of how we might distribute a brush full of paint. The paint on the brush is the total power. We might spread the paint very thinly over a large area or we might put a thick blob of paint in one spot and the rest of the area has no paint at all. The thickness of the paint at each point is the intensity at that point.

Intensity will vary as we measure it at various points in the **sound field**.

Power = energy per second (radiated by transducer)

Intensity = power crossing $1\,cm^2$ (e.g. at a sample point in the beam)

As the intensity is the energy crossing an area perpendicular to the beam axis, its value will also be proportional to the acoustic pressure squared.

i.e. $I \propto p_0^{\,2}$

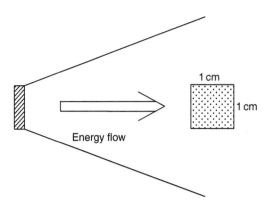

FIGURE 1.22   The relationship between power and intensity in an ultrasound beam. Intensity is the energy flowing across a unit area in the beam.

> **NOTE**
>
> Intensity is a measure of the power delivered to a point, e.g. in tissue. It is, therefore, the intensity that has most bearing on safety issues relating to medical ultrasound (see Chapter 11).

DECIBELS

It is often useful to express the change in energy, intensity, or power from one point to another as a ratio. As these changes can vary over a very large range of values, for example, the echo signal from the blood may be 10 000 less than that from a liver–fat interface; it is easier to use a logarithmic scale to compare values. The decibel scale (dB) is such a logarithmic measure. It is fully explained in Appendix 2.

A useful way to remember what the ratio is from the value in decibels is to remember

- A doubling $= 3\,\text{dB}$ and $10\times = 10\,\text{dB}$
- A negative decibel value means a fraction, so $-3\,\text{dB} =$ a half and $-10\,\text{dB} =$ a tenth.

CHAPTER 2

The Interaction of Ultrasound with Tissue

From the ultrasound transducer, pulses of ultrasound are sent into the body and the sound interacts with the structures within the tissues in a number of ways. We can look at a single pulse propagating through the body and consider what happens on its journey. What happens will affect what we see on the image and in this chapter, we will point out the clinical significance of these factors but will discuss their implications more fully in Chapter 7 on image quality and artefacts.

REFLECTION AND TRANSMISSION AT A PLANE INTERFACE

In Chapter 1 we defined acoustic impedance Z_0 and said that it was differences in acoustic impedance between two mediums that caused ultrasound to be reflected at an interface. This phenomenon is key to the formation of ultrasound images as it is ultrasound reflected back to the transducer that is detected and forms the signal from which the image is produced.

Specular Reflection

Reflection at a plane interface between two tissues with different acoustic impedances Z is called **specular reflection** from the Latin 'speculum' for a mirror.

Assuming perpendicular incidence of a plane wave on to a plane smooth interface, some of the energy in the sound wave is reflected and some will be transmitted

Ultrasound Technology for Clinical Practitioners, First Edition. Crispian Oates.
© 2023 John Wiley & Sons Ltd. Published 2023 by John Wiley & Sons Ltd.

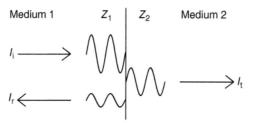

FIGURE 2.1   The intensity reflected I_r and transmitted I_t from an incident sound wave of intensity I_i at an interface between two materials with acoustic impedances Z_1 and Z_2.

across the interface and go on to deeper structures (Figure 2.1). If medium 1 has acoustic impedance Z_1 and medium 2 has acoustic impedance Z_2, then the reflected intensity I_r as a proportion if the incident sound intensity I_i is given by

$$\frac{I_r}{I_i} = \left(\frac{Z_2 - Z_1}{Z_2 + Z_1}\right)^2 = \left(\frac{1 - \dfrac{Z_1}{Z_2}}{1 + \dfrac{Z_1}{Z_2}}\right)^2$$

In other words, the ratio of $Z_2{:}Z_1$ determines the echo strength, and that depends on the density and speed of sound of the two mediums (recall $Z = \rho{\cdot}c$).

There is no energy loss at the boundary, so what is not reflected is transmitted into the second medium.

$$I_t = I_i - I_r$$

NOTES

- The order of Z_1 and Z_2 does not matter as the function is squared. If $(Z_1 - Z_2)$ is negative, squaring gives a positive value.
- If $Z_1 = Z_2$ i.e. if ρ increases and c decreases such that Z stays the same, then there will be no reflection even though the interface is between two different materials.
- If pressures are being measured, rather than intensity, then

$$\frac{P_r}{P_i} = \left(\frac{Z_2 - Z_1}{Z_2 + Z_1}\right)$$

so no squaring – and then the order of Z_1 and Z_2 does matter, as will be seen when we look at the construction of a transducer in Chapter 4.

Reflection at an Angle and Refraction

If the incident sound wave is not perpendicular to the interface, then the reflected wave will be reflected at an angle away from the incident direction of propagation. The transmitted wave will also travel in a different direction from the incident wave due to a phenomenon called **refraction** (Figure 2.2).

The angle the reflected wave travels away will be equal to the angle the incident wave arrives at but on the other side of the perpendicular line. This is expressed in the **Law of reflection** – 'The angle of reflection equals the angle of incidence'.

Reflection can produce artefacts in the image due to the misplacement of reflected targets in the image.

NOTE

A smooth interface reflecting echoes will only be seen when it is normal to the direction of propagation and the echoes come directly back to the transducer. A rough interface will be seen over a wider range of angles as sound is reflected into a range of angles. However, a rough surface will still be seen most clearly when perpendicular to the ultrasound beam (Figure 2.3).

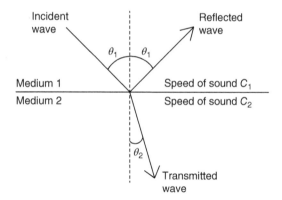

FIGURE 2.2   The angles of reflection θ_1 and refraction θ_2 of a sound wave incident at a boundary between two materials with different speeds of sound $c_2 < c_1$.

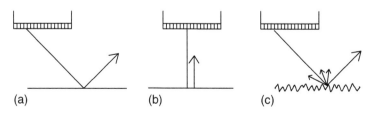

FIGURE 2.3   Showing reflection of echoes from a smooth interface when the incident wave is not perpendicular to the surface (a), compared to a perpendicular incidence (b), or a rough interface (c).

The transmitted sound is **refracted.** This means that the forward transmitted beam travels at a different angle to the line of the incident beam. The **angle of refraction** depends on the difference in the **speed of sound** between the two mediums and is described by Snell's law.

Snell's law of refraction shows that the ratio of the sine of the angles of incidence and refraction is given by the ratio of the speed of sound across the interface.

$$\frac{\sin\theta_2}{\sin\theta_1} = \frac{c_2}{c_1}$$

As shown in Figure 2.4a, when c_2 is less than c_1, the beam is refracted towards the normal of the interface and in Figure 2.4b, when c_2 is greater than c_1, the beam is refracted away from the normal of the interface.

In order to remember which way the angle changes, think of soldiers marching in formation on a hard surface moving on to a loose gravel surface (Figure 2.5). As they march on to the softer gravel surface, they start to move forward more slowly. The line of direction of the squad moves towards the normal at the change in surface. That is, crossing on to a slower speed of sound moves the direction of propagation towards the normal.

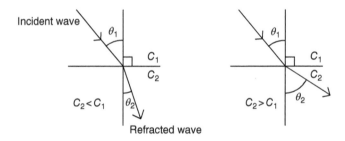

FIGURE 2.4   The angle of refraction at an interface between two materials when speed of sound (a) $c_2 < c_1$ and (b) $c_2 > c_1$.

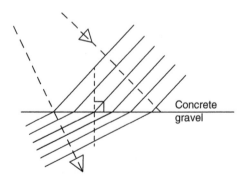

FIGURE 2.5   Illustration of the effect on lines of soldiers when marching from a hard surface on to the gravel.

In both mediums the frequency c of the sound wave is the same, so when the sound travels into a medium with a slower speed of sound, the wavelength λ of the sound waves gets shorter.

$$\text{Recall}: c = f\lambda \quad \text{i.e.} \quad \lambda = \frac{c}{f}$$

NOTE

When the angle of incidence is 90° to the interface, the wavelength will still change as the speed of sound changes and the frequency stays constant, but there will be no refraction.

REMEMBER

- The reflection coefficient involves acoustic impedance Z_1 and Z_2.
- The angle of refraction involves the speed of sound c_1 and c_2.

Refraction can produce distortions in the ultrasound image as the speed of sound is not exactly the same in all soft tissues ($c = 1540\,\text{ms}^{-1} \pm 5\%$). For example, the interface between a layer of fat and muscle may be distorted in the image due to the small difference in the speed of sound.

POOR VISUALISATION

In some patients, complex changes in the speed of sound within overlying tissue can distort the ultrasound beam to the extent that the image becomes difficult to interpret. Some types of fatty tissue consist of dense globules of fat in a matrix of less dense fat. What was a well-behaved narrow beam becomes wider and more poorly defined due to refraction. This causes poor visualisation of deeper structures in the body. It is often seen in very obese patients but is not restricted to this body type. It just depends on the type of fatty tissue. Different positioning of the probe on the skin may help, but in many cases, it is a physical limitation of the ultrasound technique and cannot be overcome.

See Chapter 7 on artefacts for illustrations of these effects. Table 2.1 gives examples of the reflected intensity at various tissue interfaces.

Summary of reflected intensity:

Soft tissues	1% or less	($< -20\,\text{dB}$)
Bone – soft tissue	50%	($\sim -3\,\text{dB}$)
Air – soft tissue	99.9%	($0\,\text{dB}$)

TABLE 2.1   Reflected Intensities at Interfaces Between Different Tissues

Interface	Reflected Intensity (%)	dB
Fat – muscle	1.1	–20
Blood (plasma – RBC)	0.005	–43
Bone – muscle	41	–4
Water – soft tissue (average)	0.23	–26
Air – soft tissue (average)	99.9	–0.004 ($\approx$0)
PZT transducer – soft tissue (average)	80	–1

NOTES

- As almost 100% of the ultrasound is reflected at an air interface, no air can be allowed between the probe and the skin. Hence, the use of gel on the skin excludes any air in the gap between the probe and the skin.
- Any gas within the body will block the path of the ultrasound beam, so we cannot see into air-filled lungs, bowel gas, etc.
- Bone reflects 50% of the ultrasound incident on it. The ultrasound that does penetrate into the bone is quickly attenuated (see below). Consequently, we cannot image the inside of the bone at higher frequencies. Some low-frequency techniques (~1 MHz) where there is some transmission through bone have been used to examine bone e.g. in osteoporosis. Where bone is thin access to deeper tissue is sometimes possible, for example, looking through the temporal bone to image cerebral blood vessels and looking at soft bone in the foetus.
- The reflectivity of blood is very low –43 dB, so moving blood appears black on ultrasound images.

SCATTERING

As ultrasound passes through tissue, what is known as **Rayleigh scattering** occurs when the sound wave encounters microscopic structures that are much smaller than a wavelength (e.g. <0.3 mm at 5 MHz).

That is scattering occurs when the target size $a < \lambda$.

As a target gets much smaller than the wavelength of the sound wave, sound energy is not directly reflected but is scattered in all directions. Only a small proportion of the scattered sound wave will return to the transducer. When the target size is

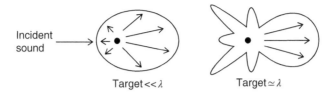

FIGURE 2.6   Diagram of examples of Rayleigh scattering from very small targets such as are seen in tissue parenchyma.

similar to the size of the wavelength, the scattering pattern is complex and depends on the precise shape and size of the scattering target (Figure 2.6).

Total Scattered Power

Total scattered power is proportional to the target size a (diameter) to the sixth power and ultrasound frequency f to the fourth power

$$\text{Scattering} \propto a^6 \cdot f^4 \quad \text{i.e.} \quad \text{it is strongly dependent on frequency.}$$

The clinical consequence of this is that frequency is an important factor in the penetration of ultrasound as it travels through tissue. High frequencies are more strongly scattered out of the ultrasound beam.

The Speckle Pattern

Ultrasound is a **coherent imaging** technique. That means that the phase of the waves can be fully specified at each point in space. In other words, at any one instant we could calculate where all of the peaks and troughs in the waves are. This may be compared with normal room lighting where each point in the room receives the light of many phases at any one time, reflected from many surfaces. Laser light is an optical example of a coherent light source.

Because ultrasound is a coherent imaging technique, scattering gives rise to the **speckle pattern** seen in images. This is the random pattern of dark-light spots at sub-millimetre level seen within tissue on the image. As shown in Figure 2.7, at the cellular level, the tissue has many small targets that are smaller than the wavelength of sound ($\lambda = 0.3$ mm at 5 MHz). Each of these targets, that lie within the length of the ultrasound pulse (e.g. 1.2 mm), will scatter the sound at the same time and send their tiny echoes back to the transducer simultaneously. These echoes have all travelled back over slightly different distances, so they arrive with slightly different phases from one another. However, the transducer will see all these echoes at once, and the signal that is detected at any one instant in time will be the sum of all the peaks and troughs of the scattered echoes it receives at that instant.

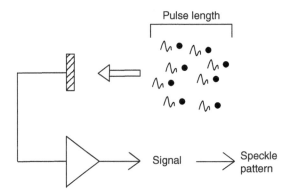

FIGURE 2.7   The origin of the speckle pattern seen in images as sound is scattered from multiple small targets within the ultrasound pulse.

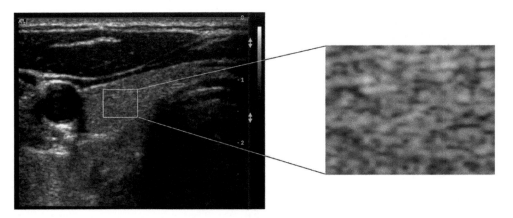

FIGURE 2.8   Image of the right thyroid showing an enlargement of the speckle pattern.

This adding together of waves is called **superposition**. One wave is superimposed on another. So, at one instant all the peaks add up to give a strong signal, and at another instant the peaks add to the troughs cancelling out to give no signal. At other points, the signal strength lies in between. The result on the image are pixels whose brightness varies in a random way to form a speckle pattern. An example of a speckle is shown in Figure 2.8.

An example of the superposition of waves is seen when watching sea waves reflect off a harbour wall. The reflected waves pass through the incoming waves and at one instant there will be a large peak as the two peaks add up and at another instant the peak of one will add to the trough of the other and for an instant the sea is flat as they cancel out.

An optical example of a speckle pattern may be seen by shining a laser pointer at a very shallow angle on a surface. A speckle pattern is seen because the laser light is coherent. In this case, it is the microscopic roughness of the surface as it reflects the light that gives the speckle we see.

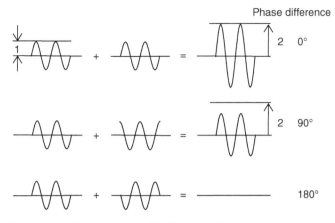

Phase difference

2   0°

2   90°

180°

FIGURE 2.9   The superposition of waves with different phases.

At a phase angle of 0°, peaks and troughs exactly add together to give twice the amplitude. At 180°, they exactly cancel out (Figure 2.9).

NOTES

- **Important** – We are not seeing the microscopic structure of the tissue in the image. What we see is a speckle pattern that depends on the frequency of ultrasound, the pulse shape, and the underlying structure of the tissue. The tissue structure is too fine to resolve as individual targets.
- Different tissues will alter the appearance of the speckle pattern, so metastases and tumours will look different to normal liver, and the speckle pattern in the liver will look different to that of the thyroid (Figure 2.10).

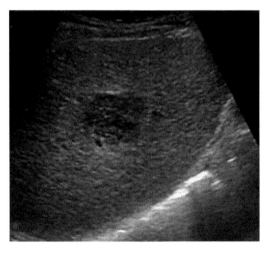

FIGURE 2.10   A hepatocellular carcinoma showing a different speckle pattern to that of the surrounding liver tissue. *Source:* Minami et al. [1], Baishideng Publishing Group Inc.

- Lower frequency ultrasound will give a coarser speckle pattern due to the longer wavelength and larger distance needed to change the phase within each pulse (Figure 2.11).

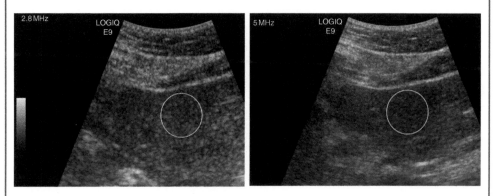

FIGURE 2.11   The liver was imaged at two frequencies: (a) 2.8 MHz and (b) 5 MHz showing a coarser speckle pattern at the lower frequency.

ATTENUATION

(**Note:** Do not confuse attenuation with acoustic impedance.)

This is the overall process by which the intensity of a wave or pulse decreases with distance from the source. It is the equivalent of the resistance of a medium to the passage of sound in that it describes the loss of energy as sound travels through the medium (cf. acoustic impedance in Chapter 1).

In addition to attenuation, there will be **boundary losses** – i.e. energy reflected away from the beam at interfaces (specular reflection).

Three mechanisms can be considered under attenuation:

1. **Absorption**
   - Some of the energy in the sound wave causes the molecules in the tissue to vibrate in a random way that is unrelated to the propagating sound wave. This is equivalent to heating the tissue. It is non-coherent vibration and is a loss of energy from the ultrasound beam.
   - For low megahertz frequencies, absorption is the predominant attenuation mechanism, accounting for 90% with scattering contributing no more than 10% [2].

2. **Scattering**
   - Energy is re-directed out of the beam by Rayleigh scattering. It is not directly converted into heat as in absorption although it will eventually be absorbed by heating tissue.
   - At low megahertz frequencies, less than 10% of the acoustic energy is scattered in soft tissue.
3. **Beam divergence**
   - As the beam travels away from the transducer, it will tend to spread out. The intensity of the beam will decrease as a result. This is not strictly an attenuation mechanism, and it depends on the beam shape (Recall: the definition of intensity – energy/unit area – as the beam spreads out intensity reduces, e.g. a pot of paint covers $1\,m^2$ very thickly or $10\,m^2$ very thinly).

> **NOTE**
>
> Absorption and scattering increase with frequency, so high-frequency ultrasound has poorer penetration than low-frequency ultrasound.

Attenuation Coefficient (α)

(Note: α does not include beam divergence.)

In a uniform medium, due to absorption and scattering, the intensity will decrease by the same fraction for every centimetre travelled. This gives rise to an exponential curve (Figure 2.12).

We could say intensity decreases by a factor of (say) 0.5 for every cm – but it soon gets hard to calculate what the intensity will be at a specific distance from the transducer.

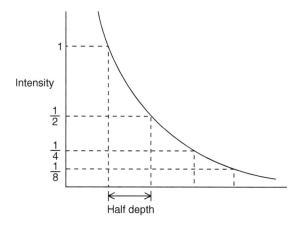

FIGURE 2.12   The change in intensity with depth due to attenuation in a uniform medium.

Therefore, we use decibels (dB) to express the attenuation coefficient. They express a ratio in a way that can be added rather than multiplied (see Appendix B).

Say, the attenuation is $-3\,\mathrm{dB\,cm^{-1}}$ (Recall: a ratio of $0.5 = -3\,\mathrm{dB}$)

Then, $8.3\,\mathrm{cm}$ of material $= 8.3 \times -3\,\mathrm{dB} = -24.9\,\mathrm{dB}$ (Easier than trying to do that using a factor of 0.5/cm)

As well as the distance travelled by the ultrasound, the attenuation coefficient will also depend on frequency – high frequencies have higher absorption and greater scattering.

For soft tissue, attenuation is proportional to frequency. Therefore, we use an attenuation coefficient α that gives the loss of sound wave amplitude in decibels per cm per MHz. Examples are shown in Table 2.2.

$$\text{Attenuation coefficient } \alpha = \mathrm{dB\,cm^{-1}\,MHz^{-1}}$$

NOTES

- This is the 'one-way' attenuation value, not 'go and return' of echoes.
- The attenuation coefficient (α) is the coefficient for change in **amplitude** of the sound wave. For a change in energy or intensity, use 2α.

Anisotropy is the phenomena of attenuation (or any other property such as speckle pattern) varying depending on the orientation of the target to the ultrasound beam or image plane. It results from the underlying structure of the tissue, for example, muscle fibres, having a longitudinal structure along their length that is

TABLE 2.2   The Attenuation Coefficient in Different Tissues

Tissue	Attenuation Coefficient, α	
Blood	0.2	
Fat	0.6	
Brain	0.8	
Liver	0.6	
Kidney	1.0	
Muscle (along fibres)	1.3	Anisotropy
(across fibres)	3.3	
Water	0.02	
Skull bone	20	

different to their cross-sectional structure. Their scattering and reflectivity, therefore, change with their orientation to the ultrasound beam.

Summary

- Attenuation varies with frequency.
- There is considerable variation between tissue types, but the average for soft tissue is $0.7\,dB\,cm^{-1}\,MHz^{-1}$.
- The bone is highly attenuating, so 50% of the pulse energy is reflected at the bone interface. Ultrasound transmitted into the bone is quickly attenuated, and we do not see inside the bone unless low frequencies are used.

THE JOURNEY OF THE ULTRASOUND PULSE

We can now summarise the journey of an ultrasound pulse as it travels through tissue (Figure 2.13). Some energy is reflected at large interfaces as specular reflection, some is scattered by structures smaller than a wavelength of sound, and some will be absorbed as heat in the tissue. When all the energy in the pulse has been dissipated, there will be no more echoes returning and the only signal in the receive circuit of the scanner will be electrical noise. This then defines the useful **depth of penetration** for the probe at that transmit power. As high-frequency ultrasound has greater attenuation, in general, higher frequency transducer probes will have poorer penetration than lower frequency probes. The depth of the target of interest will therefore influence probe choice.

USER CONTROL

Transmit Power

Transmit power is a control that the user can adjust. We could increase the transmit power to give greater penetration before the energy was dissipated but that could have safety implications, for example, physiotherapy uses high-power ultrasound to

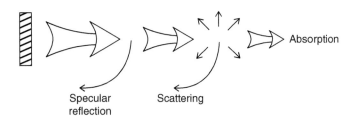

FIGURE 2.13 Diagram of the 'ultrasound journey' through tissue.

actually affect the tissue. For diagnostic imaging, we must limit the power output of the transducer, which in turn limits the depth of penetration.

The reduction in signal strength with depth, due to attenuation, can be compensated for by increasing the amplification or gain applied to the echo signal. This is fully discussed in Chapter 5.

Through its interaction with tissue, ultrasound is able to probe and give information on many physical properties of tissue, such as its speed of sound, stiffness, and attenuation. These in turn provide diagnostic windows into the tissues, shedding light, and giving information on the physiological and pathological status of the tissue. Quantitative measurements may be obtained and changes can be monitored. Later chapters will look at some of these possibilities in more detail [3].

REFERENCES

1. Minami, Y. and Kudo, M. (2010). Hepatic malignancies: correlation between sonographic findings and pathological features. *World Journal of Radiology* 28: 249–256.

2. Duck, F.A. (1990). *Physical Properties of Tissue: A Comprehensive Reference Book*, 73–135. London, UK: Academic Press.

3. Cloutier, G., Destrempes, F., Yu, F., and Tang, A. (2021). Quantitative ultrasound imaging of soft biological tissues: a primer for radiologists and medical physicists. *Insights Imaging* 12: 127–147. https://doi.org/10.1186/s13244-021-01071-w.

CHAPTER 3

Beam Shapes

In this chapter, we look at the beam shapes used for clinical ultrasound imaging. In Chapter 4, we will look at the transducers used to produce these beams.

The ultrasound beam can be thought of as the 'corridor' along which ultrasound (continuous transmission or pulses) is sent out and along which the echoes travel back to the transducer. The following discussion assumes **continuous wave** (**cw**) transmission at a single frequency. Pulses will be considered later.

If the source of a sound is a **point source** – by which we mean it is much smaller than the wavelength of the sound – then the sound will spread out in all directions (Figure 3.1). An example of a point source is a stone thrown into a pond. It sends ripples equally in all directions, the stone being smaller than the wavelength of the ripples. Another example is when you speak. The wavelength of sound in air is ~0.5 m. As this is larger than the size of your mouth, the sound spreads in all directions from your mouth and someone standing behind you can hear you.

If the source of a sound is an infinite plate moving backwards and forwards (cf. the piston), then the sound would propagate away from that plate only in the forward direction as **plane parallel waves**.

The ultrasound transducers we use fall somewhere in between these two extremes. We want narrow beams of ultrasound so that when we point our ultrasound beam at a particular target, we only get echoes from that target and nowhere else. The width of the radiating surface of the transducer is called the **aperture**. Note that for ultrasound transducer arrays, the aperture at any one time may be less than the full width of the transducer, that is, only part of the array is being used.

Ultrasound Technology for Clinical Practitioners, First Edition. Crispian Oates.

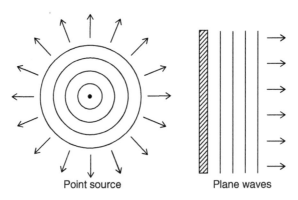

Point source   Plane waves

FIGURE 3.1   Wavefronts generated by a point source and an infinite plane source.

SIMPLE BEAM SHAPE MODEL

To a first approximation the shape of an unfocused ultrasound beam, produced by a plane (flat) transducer, consists of two distinct zones, as shown in Figure 3.2.

The **near field** or **Fresnel zone** has a uniform diameter along its length. In the **far field** or **Fraunhofer zone**, the beam spreads out in a uniform manner.

$$\text{The length of the near field is } D = \frac{a^2}{\lambda},$$

where a is the half aperture width and λ is the wavelength of sound.

The spread of the beam in the far field is

$$\sin\theta = \frac{\lambda}{a} \text{ for a square transducer}$$

or

$$\sin\theta = 0.6\frac{\lambda}{a} \text{ for a circular transducer}$$

where θ is the half angle of the spread.

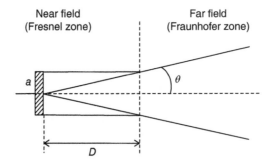

FIGURE 3.2   Diagram of the simple model of an ultrasound beam.

TABLE 3.1   Examples of Transducer Parameters for a Plane Square Transducer

λ (mm)	f (MHz)	a (mm)	D (MM)	θ
0.3	5	10	333	1.7°
0.3	5	5	83	3.4°
0.3	5	1	3	17.5°
1.5	1	10	66	8.6°
1.5	1	5	16	17.5°
1.5	1	1	0.6	Not defined

Table 3.1 shows what this means for a plane square transducer using the follow-ing values:

$$c = 1540\,\text{ms}^{-1} \qquad \text{(a)}\, f = 5\,\text{MHz} \qquad \lambda = 0.3\,\text{mm}$$
$$\text{(b)}\, f = 1\,\text{MHz} \qquad \lambda = 1.5\,\text{mm}$$

NOTE

$\sin(90°) = 1$; so, if the aperture $\leq 2\lambda$, then $\lambda/a > 1$. As $\sin\theta$ cannot be greater than 1, the source will act as a point source and sound will radiate in all directions.

SUMMARY

- A large aperture for a given wavelength λ gives a uniform beam with a long near field and a small spread in the far field.
- As the aperture approaches the size of one wavelength, the source becomes like a point source with a short near field and a widely spreading far field.
- Narrow beams require a wide aperture compared with the wavelength.
- We can reduce the aperture size and still have a narrow beam by reducing the wavelength, i.e. we increase the frequency.

In order to have narrow beams, imaging ultrasound uses high-frequency ultra-sound for a given aperture.

NOTE

Most ultrasound imaging uses wide enough apertures so that it uses the near field for imaging where the beam does not diverge over the usable depth of penetration. In this region, the beam can also be focused (see below).

The 'near field' – 'far field' picture is an approximation of the beam shape. If we plot the intensity within a real ultrasound beam from a plane (flat) transducer, we

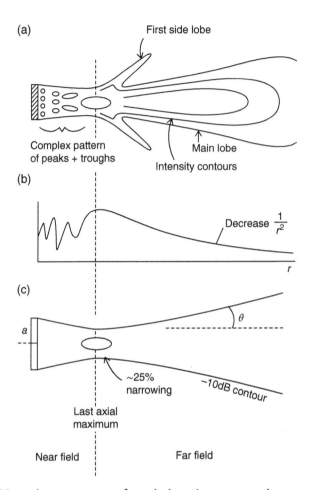

FIGURE 3.3   (a) Intensity contour map of a typical continuous wave ultrasound beam from a plane square transducer, (b) the axial intensity, and (c) comparison with the simple beam shape model.

get the diagram shown in Figure 3.3. The lines are lines of constant intensity, that is, it is an **intensity contour map** of the beam. The centre plot (b) is the **axial intensity** – along the central axis of the beam. The lower diagram (c) compares these two with the simple diagram of near field and far field we have already considered.

In the plot of the intensity changes within a real beam, we see a complex pattern of changes in the near field with a single maximum peak on the axis at the transition from the near to the far field. At this point the beam narrows by about 25%, equivalent to a natural focusing of the beam. Beyond this point, in the far field, the beam spreads out. We also see **side lobes** spreading out from the main beam just beyond the last axial maximum (near field). Echoes arising from the side lobes will also contribute to the received signal. As they are not from the main beam their signal produces **clutter** in the image, degrading the diagnostic image data. Figure 3.4 shows a beam plot from a 4 MHz plain transducer.

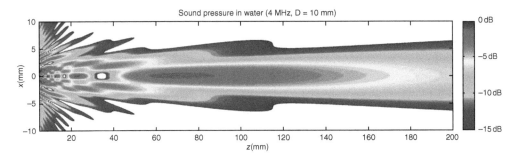

FIGURE 3.4   Beam plot in water of an unfocused ultrasound beam from a 10 mm diameter 4 MHz plane transducer.

NOTE

The intensity variations are less pronounced for short pulses than for cw ultrasound because short pulses contain a range of frequencies and therefore tend to smooth out the fine variations in intensity.

The concept of a beam applies to both transmission and reception. In other words, received echoes travel back to the transducer along a beam with similar properties to the transmitted beam we are considering. Transmission and reception are therefore reciprocal. However, in a real scanner, the receive beam does not have to be the same as the transmitted beam. We can set up a different beam shape for reception than that used for transmission (see Chapter 6). The net beam shape for the transmit-echo round trip will then be the transmit and receive beams shapes multiplied together.

HUYGEN'S WAVELET MODEL AND DIFFRACTION

A useful way of understanding how the beam shapes we see arise is to use Huygen's wavelet model (Figure 3.5). A plane transducer aperture can be modelled as many point sources next to each other. These are known as **Huygen's sources** and their waves as **Huygen's wavelets**. The Huygen's wavelets from each point source spread out spherically from that source (Figure 3.5a). As shown for the transducer with a limited aperture (Figure 3.5b), the wavelets from Huygen's sources across the aperture add by superposition so as to give the expected plane wave in front of the 'piston', but at the edges they spread out sideways with no more sources to interact with. The pattern so produced is called the **diffraction pattern** for that source or aperture. The interaction of the wavelets with one another, through superposition, is known as **interference**. In general, the shape of the beam we get from a transducer (if there is no reflection, refraction or attenuation) is the result of the diffraction pattern from the source or aperture of the transducer.

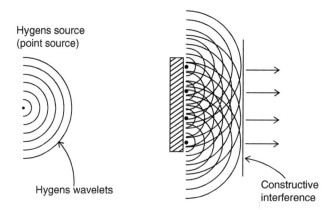

Hygens source
(point source)

Hygens wavelets

Constructive
interference

FIGURE 3.5   Huygen's wavelets for (a) a point source and (b) a plane transducer aperture.

Where a peak adds to a peak to give larger intensity, it is known as **constructive interference** and where a peak and trough cancel each other out, it is known as **destructive interference**.

Using Huygen's wavelet model enables us to see how the various beam shapes used in ultrasound arise. A more detailed explanation of how the shape of the beam we see arises is given in Appendix C.

FOCUSING

We have seen that a plane unfocused transducer has some narrowing of the beam at the last axial maximum of ~25%. This is a natural focusing.

Consider the idealised geometrical arrangement of a spherical bowl-shaped transducer shown in Figure 3.6.

Huygen's wavelets generated from across the transducer aperture constructively interfere to produce converging wavefronts that all arrive at the same time at F. So, F sees a high intensity just at that point. At other field points in that range, there will be destructive interference and low intensity or no sound wave at all.

In a real physical situation, because of diffraction at the edges of the aperture, a spherically curved transducer will produce a narrow beam at a focal point, but the beam will still have a finite width as shown in Figure 3.7. The use of an **acoustic lens**, made of plastic or rubber, similar to a glass lens in optics, can also be used to focus an ultrasound beam.

The width of the beam at the focus is $\quad W = F\dfrac{\lambda}{a}$,

where F is the **focal length**, a is the half aperture width, and λ is the wavelength of sound. Figure 3.8 shows a focused beam from a linear array.

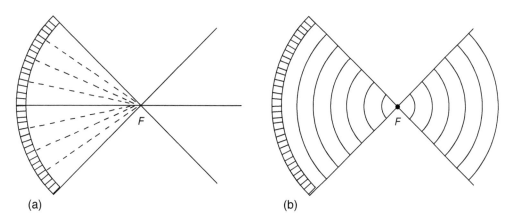

FIGURE 3.6 Idealised geometrical focus from a spherical transducer shown as (a) geometrical rays and (b) converging wavefronts.

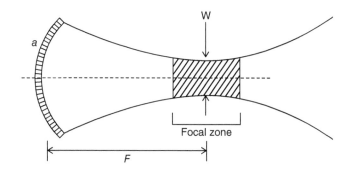

FIGURE 3.7 A spherical transducer showing the definition of the focal zone and focal width W for a focal depth F.

NOTES

- The **focal zone** or axial length of the focus is defined as the region where the beamwidth is $\leq 2W$.
- For a focused transducer, the focal point equates to the far field of an unfocused transducer.
- A large aperture gives a narrow **focal width** just as a large aperture gives a small beam spread ($\sin \theta$) for an unfocused far field.
- The **focal length** (distance from the transducer to the focus) must be less than the near field length of the unfocused transducer. In other words, you can only focus the beam in the near field. By using wide aperture transducers, the near field will be longer and focusing can be used throughout the useful range of the image. Ultrasound beams from clinical imaging probes generally use the near field region for imaging (Figure 3.8).

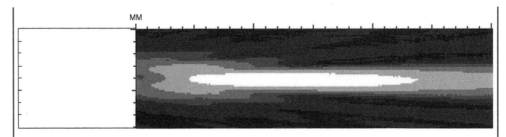

FIGURE 3.8   Beam plot of a Doppler pulse from a linear array with a 2.9-cm focal depth. The first centimetre nearest the transducer is not shown. *Source:* Courtesy of Barry Ward.

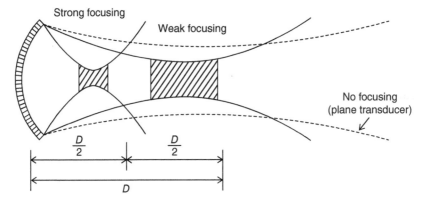

FIGURE 3.9   The difference in beam shape with strong and weak focusing.

The **strength of focus** may be defined as the ratio of the aperture width a to the width of the beam at the focus W.

$$\frac{a}{W} = \frac{a^2}{F\lambda} = \frac{D_{unfocused}}{F} \quad \text{where } D \text{ is the near field length.}$$

Where the focal length is less than half D we have strong focusing, and where it is greater than half D we have weak focusing (Figure 3.9).

NOTES
- For weak focusing, the focal zone is longer than that for strong focusing.
- There is an improvement in beamwidth at the focal zone but beyond that the beam spreads out *more* than it would for an unfocused beam. This has implications for visualising tissue in front of or beyond the focal zone. The target of interest should be in the focal zone (see below).
- To have strong focusing at depth, you need a wide aperture.

BEAM FORMING WITH TRANSDUCER ARRAYS

By using arrays of elements to build a transducer, all aspects of beamforming can be controlled electronically. This includes controlling the aperture width and position, steering and focusing. These can all be changed in **real time**, that is, as the image is acquired.

As we consider simple arrays, it is important to note that the ultrasound beam will have a beam shape in the **scan plane** or **image plane,** that is determined by how the elements of the array are driven, and a beam shape in the **elevation plane** that will be the beam shape of a single transducer element (Figure 3.10). It is this elevation plane beam shape that will determine the **slice thickness** of our image. It is important to recognise that the image produced is not an infinitely thin cut through the tissue. The slice thickness can be optimised by using an **acoustic lens** in the front of the transducer so as to give weak focusing throughout the elevation plane.

It is easiest to begin by considering a **linear array** that will produce a **linear scan image,** as we saw in the introduction (Figure 0.1a).

Figure 3.11 shows a small array of four elements.

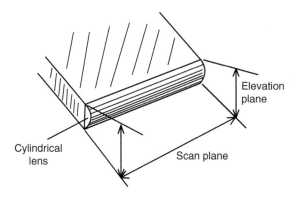

FIGURE 3.10   Definition of scan plane and elevation plane for a linear array transducer.

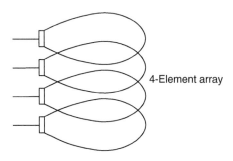

FIGURE 3.11   A 4-element array with beams from each element indicated.

NOTES

- Each element in the array has its own beam shape, which is the beam shape produced by a single element transducer as discussed above.
- The array elements are small so the beam shape from each element is wide and spreads over a wide far field angle.
- Wavefronts from all the elements combine by interference to give the wide aperture beam shape we want to make from the array as shown in Figure 3.12.

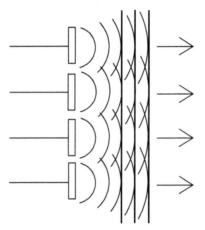

FIGURE 3.12   Wavelets from a 4-element array showing the alignment of wavefronts to give plane waves from the whole array.

The strongest reinforcement of intensity is the tangent of all the wavefronts, i.e. straight ahead. This looks like the beam shape we would get from one large aperture. However, in an array, although each element will have its own beam shape that approximates a point source, there are no waves being generated from the gaps between the elements.

Grating Lobes

Analogously to the formation of side lobes (see Appendix C), an array can produce **grating lobes** due to diffraction effects between the individual elements. A grating lobe is a directional lobe of increased intensity where there is positive reinforcement of waves from individual elements in the array.

Grating lobes arise because the array uses independent elements with gaps between them (Figure 3.13). Along the line of the first grating lobe, the wavelets from each of the elements are in phase as the path difference is exactly one wavelength. They, therefore, reinforce each other and we have a grating lobe (a secondary beam) formed away from the main beam axis. In principle, there could also be a second grating lobe when the path difference is exactly two wavelengths. The angle of the

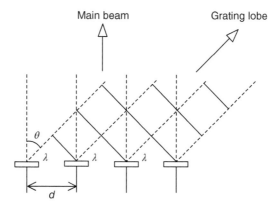

FIGURE 3.13 Diagram showing the origin of a grating lobe at an angle giving one wavelength difference in path length.

grating lobe depends on the spacing of the transducer elements. The wider apart the elements, the closer the grating lobe is to the main beam. The intensity of the grating lobe is lower than the intensity of the main beam.

The angle of the grating lobe from the main lobe is given by

$$\sin\theta = \frac{\lambda}{d} \quad \text{where } d \text{ is the distance between elements.}$$

If the distance between the elements is equal to or less than a wavelength, i.e. if $d \le \lambda$, then $\sin^{-1}(\theta)$ is undefined as the value is >1 and $\sin(90°) = 1$. Since an angle of 90° would be along the line of the array, there will be no grating lobe for the beam. In other words, the elements are then close enough together to make the array behave like a single transducer. In general, linear and curvilinear arrays used for medical imaging will have element spacing $>\lambda$, so there will be grating lobes (Figure 3.14).

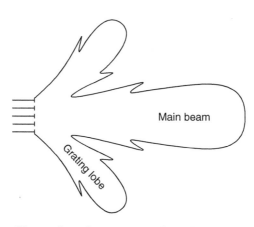

FIGURE 3.14 The overall beam shape for an array with grating lobes.

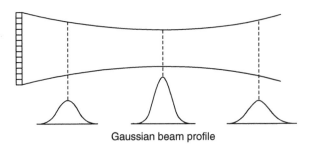

Gaussian beam profile

FIGURE 3.15   Using apodization, the beam can be given a Gaussian profile which remains constant throughout the length of the beam.

The exception is for **phased arrays** which have fine element spacing of <1λ (see Chapter 5).

Apodization

For each individual element in the array, the phase and amplitude of the signal driving that element during transmit can be controlled.

Shaping the beam across the active aperture by tailoring the amplitude to each element is called **apodization**. By a careful choice of amplitudes, the sidelobes in the beam can be greatly reduced. In particular, a beam cross-section profile with a **Gaussian shape** has reduced side lobes and maintains its shape along the length of the beam (Figure 3.15). The beamwidth is slightly wider than for a non-apodized beam, but the benefits of a more uniform beam at all depths are useful and some scanners use **Gaussian beams**.

BEAM STEERING

It is useful to be able to steer a beam electronically using an array. This can be done using time delays to alter the phase and hence the position of the wavefronts (Figure 3.16).

For example, by delaying the phase of the wavefront by Δt between adjacent elements, we can cause the wavefronts to 'add up' along a direction at an angle to the straight ahead beam. When the peak transmitted from the first element has moved out by s in the direction we want to send the beam; the peak from the next element which has been delayed by Δt is just leaving its element, and the two peaks are in phase and constructively interfere along the direction we want to steer the beam.

$$s = c \cdot \Delta t \quad \left(\text{Recall}: \text{velocity} = \text{distance} / \text{time}\right)$$

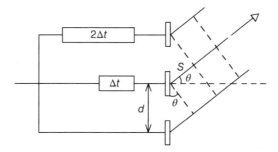

FIGURE 3.16 Diagram of a steered transmitted beam using time delays to delay the phase of wavefronts from each element.

So,

$$\sin\theta = c\frac{\Delta t}{d}$$

where d is the distance between elements, and c is the speed of sound in tissue.

Therefore, to steer the beam by 30° (sin (30°) = 0.5) on an array with element spacing $d = 0.5\,mm$, $c = 1540\,ms^{-1}$

$$\Delta t = d\frac{\sin\theta}{c}$$ so, the phase delay between elements needs to be $\Delta t = 0.11\,\mu s$.

Using the same time delays for the returning echoes, the wavefronts from the steered direction will be brought back together and be in phase with each other. Returning echoes reaching each element are delayed by Δt and then summed so phases of echoes arriving from the steered direction sum constructively. Those from other directions sum destructively and are not detected (Figure 3.17). The transducer will then have greatest receive sensitivity to echoes arising from the steered direction. This technique is called delay and sum beamforming.

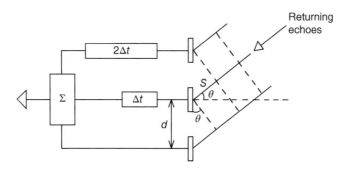

FIGURE 3.17 Diagram showing how a received steered beam is detected with time delays and summation of signals from each element.

Beam Shape and Beam Steering

> **NOTES**
>
> - The main lobe has the greatest intensity/sensitivity when it is aligned with the main axis of the array; in other words, when it is steered forward. This has an important implication when looking for weak targets with Doppler and colour Doppler – the greatest sensitivity is obtained when the beam is steered forward.
> - As the beam is steered away from the main axis, the main beam gets weaker and the grating lobe becomes stronger as it approaches the forward position (Figure 3.18).
> - **Phased arrays** (see Chapter 6) scan the image plane by steering the beam through a wide angle. In order to completely suppress the grating lobes, their elements are spaced $<\lambda/2$ apart so that even with steering up to 90°, $\sin^{-1}(\theta)$ is >90° for the grating lobe; in other words, it does not exist.

Beam steering also has another effect on image quality. As the beam is steered, the effective aperture of the transducer is reduced, so the beam will spread more and resolution will be reduced (Figure 3.19) (see below for more on resolution).

Figure 3.20 shows a beam plot of a steered beam from a linear array.

ELECTRONIC FOCUSING

A physical lens placed in front of a light source or a sound source causes delays in the wavefronts so that they bend and converge to form a focus. In a similar manner to beam steering, the same effect may be produced using time delays between the

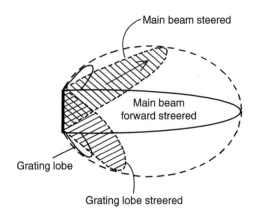

FIGURE 3.18   Diagram showing how steering the beam decreases main beam sensitivity and enhances the grating lobe signal.

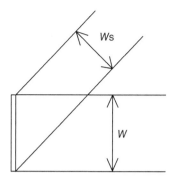

FIGURE 3.19 The effect of beam steering on the effective aperture of the transducer.

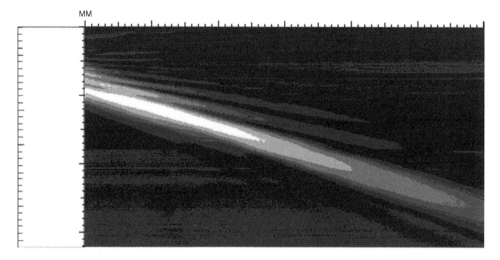

FIGURE 3.20 Beam plot of a steered Doppler pulse from a linear array with a 5.6 cm focal depth. Note the side lobes appearing towards the straight ahead direction. Source: Courtesy of Barry Ward.

elements of a transducer array. The timings between the elements are chosen so that the transmitted wavefront from adjacent elements reinforces along a curved line which converges on a field point, equivalent to focusing on that point (Figure 3.21). The effect is as though the beam had been produced by a spherically curved transducer (see Figure 3.6).

As for steering, by delaying the wavefronts arriving at the transducer with the same time delays as were used in transmit, the transducer will have the greatest sensitivity for echoes arising from the focal point. After passing through the time delays, the wavefronts are summed. The delays bring the phase from each element together, so they reinforce each other for that focal point.

By changing the time delays, the **depth of the focus** can be controlled. By changing the width of the aperture (number of elements used), the **focal width** can be controlled.

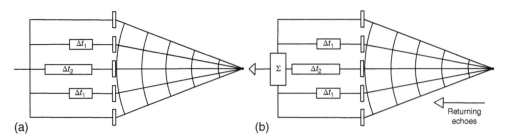

(a)   (b)

FIGURE 3.21   Diagram showing the use of time delays to produce the converging wavefronts of a focused transducer (a) in transmitting and (b) in receiving.

RESOLUTION

Ideally, we would like to image organs with very fine detail shown in the image so we could detect very small but clinically significant changes. However, there is a limit on how small a target we can clearly image. The **resolution** of an imaging modality is the minimum size of the target that can be imaged.

As a pulse travels down the ultrasound beam, it typically has a teardrop shape in space, shaped by the image plane (x-axis) and elevation plane (y-axis) beam shapes, and the pulse length (Figure 3.22).

> **DEFINITION**
>
> **Resolution** is the ability to resolve two targets as separate targets – rather than one blob on the image, as shown in Figure 3.23.

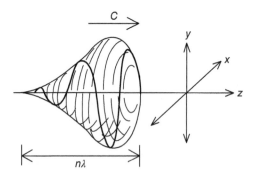

FIGURE 3.22   The 'teardrop' shape of an ultrasound pulse in space.

Target   Just resolved   Not resolved

FIGURE 3.23   For two small targets closely spaced (a), they are just resolved in the image if they are seen as two targets (b) and not resolved if they appear as a single target (c).

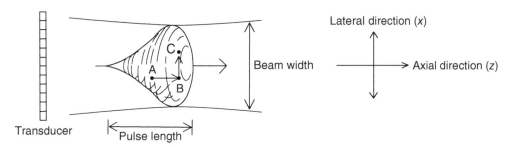

FIGURE 3.24   Location of points (A, B, C) within the pulse used to define image resolution.

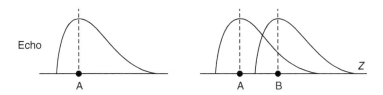

FIGURE 3.25   Resolving points A and B along the beam axis.

We need to consider the resolution along the axis of the ultrasound beam, **axial resolution,** and in the transverse image plane, **lateral resolution** (Figure 3.24).

Axial Resolution

Resolution along the direction of propagation A-B (Figure 3.25).

A and B are two locations in the tissue that are both inside the pulse at the same time, therefore both send back echoes simultaneously. As the pulse passes A in the Z (time) direction, the echo from A maps out the pulse length shape on the image, i.e. not a single point but a blob. Similarly for B.

A and B will just be separated if B is just coming into the pulse as A is just leaving the pulse. This will be the case if A and B are half a pulse length apart, so

$$\textbf{Axial resolution, } A - B = \frac{n\lambda}{2} \quad \text{where } n \text{ is the number of wavelengths } \lambda \text{ in the pulse.}$$

Any targets closer than this will *not* be resolved as separate targets.
To improve the axial resolution, we would need

- Fewer cycles = shorter pulse

or

- A higher frequency = shorter wavelength, λ

Lateral Resolution

Resolution across the beam in the image planes B-C (Figure 3.26).

B and C are two locations across the beamwidth that are both inside the pulse at the same time. The ability to separate B and C depends on how wide the beam is, for

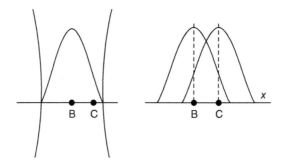

FIGURE 3.26   Resolving points B and C across the beamwidth.

example, at the focus the width is small. As the ultrasound beam is swept across B and C to form the image, the echo from B maps out the beamwidth profile on the image producing a 'blob', not a single point. Similarly for point C.

As for axial resolution, B and C will just be separated if the distance B-C is half the beamwidth. In other words, B just leaving the beam as C just enters it.

Any targets closer than this will not be resolved as separate targets.

To improve the lateral resolution, we would need a narrower beam. We could do this by:

- Using a higher frequency
  - Recall: for a plane transducer the near field distance $D = a^2/\lambda = a^2f/c$, so a higher frequency means we can use a narrower aperture a with a narrower beamwidth.

or

- Stronger focusing

In order to look at the resolution, we have considered three targets A, B, and C that the pulse shape sample volume passes. Another way of looking at it is to see that everything inside the sample volume of the pulse adds together to give the echo we detect and so all the information from within the sample volume of tissue is superimposed in one echo. In other words, that information is degraded. It is these echoes arising from within the sample volume, i.e. pulse length/2 or beamwidth/2 that give rise to the unresolved **speckle pattern** in the image due to **Rayleigh scattering** (see Chapter 2). An example of the high resolution that can be achieved using very high frequencies is seen in Figure 3.27.

NOTE

Lateral resolution will be poorer if line density is insufficient to cover the target (and potential targets will be missed). For sector shape imaging line density increases with distance (Figure 3.28). Line density will be discussed further in Chapter 7.

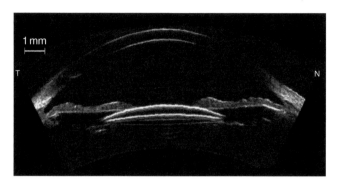

FIGURE 3.27 Anterior chamber of the eye with a prosthetic intraocular lens. An example of very high-frequency imaging at 50 MHz showing the high resolution of <60 μm that can be achieved with very high frequencies [1].

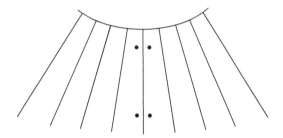

FIGURE 3.28 Diagram showing how line density changes with depth for a sector scan.

SUMMARY

In order to have both good axial resolution and good lateral resolution, we need to use high frequencies. High-frequency ultrasound probes will always give better resolution than lower frequency probes.

Resolution Versus Penetration

We have seen that in order to have a fine resolution, a high frequency is needed. However, in Chapter 2, we saw that higher frequencies suffer from higher attenuation due to greater scattering and absorption and so have less penetration. There is therefore a trade-off between **resolution** and **penetration**.

NOTES

- Use the highest frequency probe that allows visualisation at the depth of the target of interest.
- Some scanners have a control that lets the user trade between resolution and penetration. This control alters the frequency of the transmitted pulse within the bandwidth of the transducer.

CLUTTER

The presence of echoes arising from side lobes, grating lobes, and multiple reflections from within the body all contribute to the signal that forms the image. Such echoes do not arrive at the transducer from the main beam and the information from them is spurious. The term used to describe the degradation of the image caused by these echoes is '**clutter**'. A number of techniques, such as using Gaussian beams, have been used to reduce the effect of clutter in the image.

REFERENCE

1. Coleman, D.J. and Silverman, R.H. (2006). Explaining the current role of high frequency ultrasound in ophthalmic diagnosis. *Expert Review of Ophthalmology* 1: 63–76.

CHAPTER 4

The Ultrasound Probe

Within medical ultrasound, the terms '**ultrasound probe**' and '**ultrasound transducer**' are often used interchangeably. Strictly speaking, the transducer is the device within the probe that produces the ultrasound and the probe is the whole assembly that the user holds and manipulates to produce the images they want.

A **transducer** is any device that converts energy from one form to another. As seen in the first chapter, a sound wave is produced if we have a mechanical vibrator such as a piston or a loudspeaker. A loudspeaker is a transducer converting electrical energy into mechanical energy – the sound wave.

In order to use ultrasound to produce medical images, we need to convert an electrical signal into an ultrasonic sound wave and then to convert mechanical movement produced by the returning echoes into an electrical signal that can be processed to form the image.

Figure 4.1 shows the basic layout of an ultrasound probe as used in medical ultrasound. We look at each of the components in turn.

THE TRANSDUCER

Most medical ultrasound transducers use the **piezoelectric effect** to convert electrical energy to sound energy and then sound energy from echoes back to electrical energy.

Piezoelectric Effect

Some materials show the property that when a voltage is applied across them, they either shrink or expand depending on the polarity of the voltage (Figure 4.2). The

Ultrasound Technology for Clinical Practitioners, First Edition. Crispian Oates.

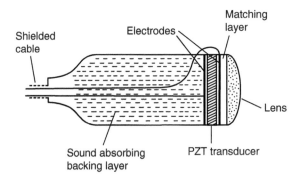

FIGURE 4.1   The basic elements of an ultrasound transducer.

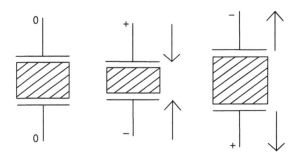

FIGURE 4.2   The voltages produced when a piezoelectric material expands and contracts.

phenomenon is called the **piezoelectric effect**. It also works the other way around. If the material is squeezed or caused to expand, a voltage forms across the opposite surfaces of the sample. The voltage is proportional to the degree of deformation of the material. Quartz is an example of a naturally occurring piezoelectric material. This forms the basis of the timer in a quartz watch, when a piece of quartz is made to resonate at a high frequency. This produces an alternating voltage that regulates the clock, as a pendulum does in a grandfather clock.

The piezoelectric effect is ideal for making **ultrasound transducers** as piezoelectric materials can be made to resonate at megahertz ultrasonic frequencies. They can be driven by applied voltages to transmit ultrasound (act as a piston on the surrounding medium) and when vibrated by the returning echoes, they produce an alternating voltage that forms the received signal for processing to form an image.

Most modern medical ultrasound transducers use a manufactured ceramic called lead zirconate titanate (or **PZT**), which can be formed to any shape required. Another piezoelectric material sometimes used is **PVDF** (polyvinylidene fluoride), which is a polymer.

Manufacture of Piezoelectric PZT Transducers

At a molecular level, untreated PZT has a domain structure in which each domain has a positive end and a negative end. These are randomly oriented. If a voltage is applied, the domains line up and the PZT sample expands by electrostriction,

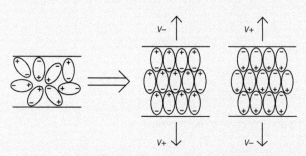

FIGURE 4.3   The behaviour of unpolarised PZT when a voltage is applied before it is made piezoelectric. Unpolarised PZT (a) expands by electrostriction as domains line up (b).

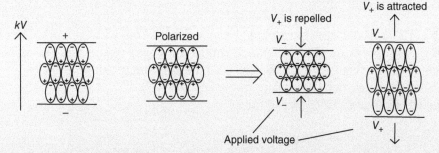

FIGURE 4.4   PZT is polarised by cooling under an applied voltage (a). It then exhibits the piezoelectric effect (b).

whichever way the voltage polarity is applied as shown in Figure 4.3.

In order to make it piezoelectric a high voltage is applied whilst the sample is heated above a critical temperature called the **Curie temperature**. The domains line up under the applied voltage (Figure 4.4). The sample is then cooled whilst the voltage is maintained and the polarisation of the domains is frozen into the sample. The sample is then piezoelectric and further application of voltages will produce the expansion and shrinking seen in piezoelectric materials.

Single Crystal Technology

Fabricating a conventional PZT transducer using the method described in the box results in a transducer in which, typically, only 70% the material is polarised. More recently, a significant improvement in transducer efficiency has been obtained by growing a single crystal of the piezoelectric material. This produces a transducer with fewer defects and lower losses which when polarised produces a near perfect align-ment of poles. Effectively, the whole transducer has a single domain for polarisation.

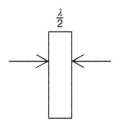

FIGURE 4.5   Half wave thickness of a transducer.

This improves efficiency by 85% over conventional PZT transducers allowing greater sensitivity and improved visualisation in difficult scanning situations.

Resonant Frequency

A slab of PZT will vibrate at its resonant frequency when 'hit' with a voltage spike, like pinging an empty wine glass to make it ring as shown in Figure 4.5. The resonant frequency will be that for which the slab is a half wavelength thick.

For example, we need a resonant frequency of 5 MHz

Speed of sound in PZT, $c = 3200 \, \text{m·s}^{-1}$

$$\text{So, } \lambda = \frac{c}{f} = \frac{3200}{5 \times 10^6} = 0.64 \, \text{mm} \quad \text{so, slab thickness, } \lambda / 2 = 0.32 \, \text{mm}$$

The resonant frequency chosen will be the **centre frequency** for the probe.

Thin metal **electrodes** are deposited on each side of the PZT slab to apply the transmit voltages and pick up the receive signal. The signal is fed through a shielded cable to avoid interference from other signals.

BACKING LAYER

We have previously seen that in order to have good resolution we need very short pulses. Therefore, we need to stop the transducer from ringing when the voltage spike is applied like pinching the edge of a wine glass whilst pinging it with the other hand. This is called **damping** (Figure 4.6). Within the ultrasound probe, damping is achieved by putting a dense sound absorbing material against the transducer forming a backing layer in the probe. It is the effect of the backing layer that enables pulses to be short and the transducer wideband in its frequency range.

NOTE

Because the backing absorbs some of the sound energy and some sound is reflected back into the transducer at the air–skin interface, the transducer can warm up. This is known as **self-heating** and is a potential hazard when placing the transducer on the skin.

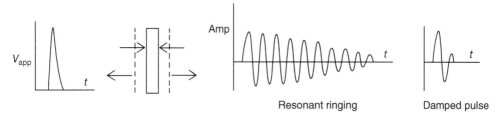

FIGURE 4.6   The process of impulsing a transducer element with a voltage spike V_{app}, showing resonant and damped ringing.

Recent improvements in the technology and design of backing layers have improved transducer efficiency and reduced the self-heating problem.

MATCHING LAYER

PZT is a hard, rigid ceramic material with a high impedance and skin is a relatively low impedance material. Recalling Chapter 2, there will, therefore, be a large reflection of ultrasound at the interface between the transducer and the body, so potentially very little sound enters the body. This problem can be overcome to a large extent by the use of a **matching layer**. This is a layer of a material whose impedance lies somewhere in between the impedance of the transducer and that of the body. It is the same feature as the anti-reflective bloom seen on the front of a camera lens or binoculars. The coloured bloom is a matching layer that allows the light to pass in or out of the lens without being reflected back into the lens at the glass–air interface.

How Does the Matching Layer Work?

Recall: Reflected intensity

$$\frac{I_r}{I_i} = \left(\frac{Z_2 - Z_1}{Z_2 + Z_1} \right)^2$$

The amplitude is proportional to the square root of intensity, i.e.

$$\frac{A_r}{A_i} = \left(\frac{Z_2 - Z_1}{Z_2 + Z_1} \right)$$

Looking at the equation for **amplitude** we see that the order of Z_2 and Z_1 matters as there is a minus sign there. If we go from a low impedance to a high impedance, the result is positive and if we go from a high impedance to a low impedance the result is negative. What this means in terms of the amplitude of

(continued)

(*continued*)

the reflected wave is that where the result is negative, there will be a **phase inversion** of the sound wave. What was positive going becomes negative going, i.e. 180° phase shift. This will happen going into the skin as the skin has a low impedance compared with the PZT.

The matching layer is made of a material whose impedance lies between that of the PZT and the body and is a quarter of a wavelength thick. It is therefore also known as **quarter wave matching** or **impedance matching.**

The way the matching layer works is shown in Figure 4.7. Any reflected sound at the body interface will have its phase inverted, i.e. a positive peak becomes a trough = $\lambda/2$ change. The sound wave then travels back through the quarter-wave layer = $\lambda/4$ change in phase due to a quarter of a wavelength in distance. It then reflects at the higher impedance of the PZT, so no phase change, and then travels back across the quarter-wave layer to the body to give another $\lambda/4$ change in phase.

So, the total body–PZT–body reflection path = $\lambda/2 + \lambda/4 + \lambda/4 = 1\lambda$

This means that the second, reflected, sound wave T_2 transmitted into the body is in-phase with the first transmitted wavefront T_1 and reinforces it. All multiple reflections will be similarly in-phase and at that one frequency there will be almost perfect transmission of the sound into the body.

The matching layer is made of a material whose impedance is the geometric mean of the impedances on either side, i.e.

$$Z_{ML} = \sqrt{Z_{body} \times Z_{PZT}}$$

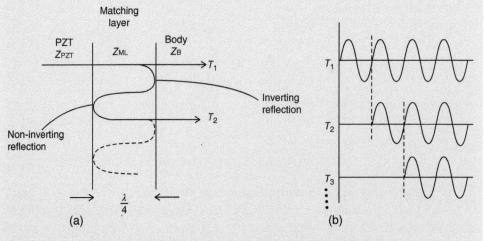

(a)   (b)

FIGURE 4.7   The phase changes occurring in a quarter wave matching layer.

NOTES

1. Quarter wave matching will only work really well for one frequency

$$f = \frac{\lambda_{ML}}{c_{ML}}$$

2. **Multiple Matching Layers**: With each layer having an impedance of the geometric mean of those on either side, will match over a wider range of frequencies such as are contained in a short pulse of sound. The effect is then closer to a gradual transition from Z_{PZT} to Z_{body}. This is then like multi-coated lenses on a camera or the bell on a trumpet matching a small tube to the open air. There is a more gradual transition from a high to a low impedance.

FRONT FACE LENS

On the front face of the transducer probe is an acoustic lens made of a soft rubber with an impedance close to that of soft tissue. In the case of a simple array probe, the lens will be a cylindrical lens across the front face of the probe giving weak focusing in the elevation plane, to minimise slice thickness, but having no effect in the scan plane. In addition to giving some focusing in the elevation plane, the front face forms a hygienic, insulating barrier to the probe, so it can be safely cleaned (see Figure 4.12).

WIDEBAND TRANSDUCERS

Advantages:

- Very short pulses are possible, giving good resolution.
- It is possible to drive the transducer with complex or 'special' pulse shapes.
- Harmonic imaging using higher frequencies is possible.
- The same transducer can use different frequencies for imaging and Doppler.

To make a transducer wideband requires the following:

- Transducer element thickness chosen to be at the middle of the wideband range.
- Efficient damping in the backing layer.
- Very good matching to body impedance.

CONSTRUCTION OF AN ARRAY

Most ultrasound probes used in medical imaging consist of an array of transducer elements aligned next to each other (Figure 4.8). They may be fabricated from a single slab of PZT etched to produce the individual elements, but the principle components of their construction are the same as for a single element transducer, as described above.

NOTES

- For a phased array, the element spacing is <λ/2 so that there are no grating lobes formed when the beam is steered through large angles.
- The array may be 'bent' to form a curvilinear array.
- The development of the technology used in constructing ultrasound probes has used new materials and techniques to improve their weight, sensitivity, and resolution. For example, each element may be further split, a method called **sub-dicing**. The gaps produced are filled with a low impedance material thereby reducing the effective impedance of the PZT element. This improves signal transmission and reception into the body.

CMUT TECHNOLOGY

More recently, it has become possible to construct transducers that operate using a varying capacitance between electrodes rather than using the piezoelectric effect of a solid material [1]. The device is called a **capacitive micromachined ultrasound transducer** or **CMUT** (Figure 4.9).

Cavities are micromachined into a silicon substrate that will resonate at ultrasound frequencies (Figure 4.10). A thin metalised membrane is suspended over the cavity. The membrane and the silicon substrate then form the electrodes, and a voltage bias is applied between them. If an alternating driving voltage is then applied on top of the bias, the changing capacitance of the cavity will cause the membrane to

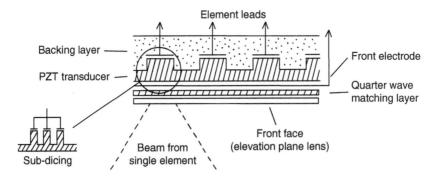

FIGURE 4.8   The basic structure of a piezoelectric transducer array.

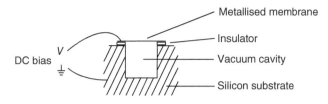

FIGURE 4.9 Diagram of the main components of a CMUT transducer element.

FIGURE 4.10 A scanning electron microscopy image of a CMUT silicon back plate showing a grid of the resonant cavities. *Source:* Cianci et al. [2]/with permission of Springer Nature.

vibrate and transmit an ultrasound wave. When the membrane is caused to vibrate by a returning echo, the capacitance of the cavity will change and an alternating voltage will be generated forming the detected signal in reception. The resonant frequency depends on the size of the cavity and the stiffness of the membrane. Many of these capacitor cells may be connected in parallel to implement transducers and arrays with different shapes and sizes and the required wide frequency response (Figure 4.11).

The surface is covered with a thin layer of elastic polymer to provide a hygienic insulating surface. A very large frequency arrange is achievable, e.g. 2–22 MHz within one probe. The acoustic impedance of the vibrating membrane is close to that of tissue, so no matching layer is required for efficient transmission into the body. A typical cavity size is 0.1–0.2 μm.

CMUT devices have a number of advantages. These are as follows:

- It is easier to construct 2-D arrays with CMUT technology using standard microelectronic manufacturing technology.
- Large numbers of CMUTs in an array enable very wide bandwidths to be achieved.

FIGURE 4.11   A portion of a CMUT array of rectangular elements made of many CMUT units (pitch: 250 μm × 12 mm). The electrode connections to each element are seen at the top. *Source:* Cianci et al. [2]/with permission of Springer Nature.

- The wide bandwidth makes high frequency operation easier to achieve, e.g. for harmonic or very high frequency – high resolution imaging.
- CMUTs have high sensitivity with low noise.
- No matching layer is required.
- Silicon technology integrates well with other electronic circuitry

1-D, 1.5-D, AND 2-D ARRAYS

The basic **linear array** is a one-dimensional array imaging in a single scan plane (Figure 4.12).

A **curvilinear array** is a linear array bent round a circular arc to give a sector shaped scan (Figure 4.13). The construction of a **phased array** is similar to that of a linear array but with very close spacing of the elements so that they are <1λ apart.

The cylindrical lens shown on Figure 4.12 gives weak focusing in the elevation plane to reduce slice thickness. This will give a general reduction in slice thickness, particularly at the focal distance of the lens. Further improvement in slice thickness has been achieved in two ways: the Hanafy lens and the 1.5-D array.

The Hanafy Lens

The Hanafy lens uses transducer elements that vary in thickness in the *y*-direction to give improved slice thickness throughout the image rage (Figure 4.14). As it is a

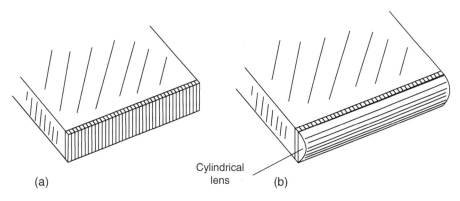

(a)

Cylindrical lens

(b)

FIGURE 4.12   Diagram of a linear array showing the elements and front face cylindrical lens.

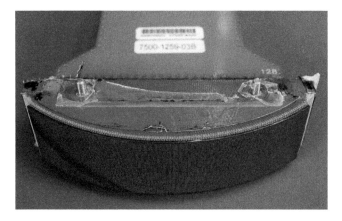

FIGURE 4.13   Showing the construction of a Philips C5-2 128 element curved array transducer inside the casing, behind the front lens facia. *Source:* Binarysequence [3]/Wikimedia Commons/ CC BY-SA 4.0.

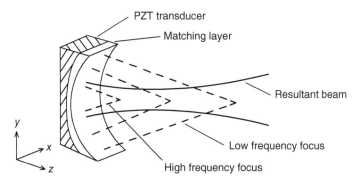

FIGURE 4.14   Diagram showing the focusing in the elevation plane of a Hanafy lens.

curved transducer, it will focus the ultrasound. Where the element is thin at its centre, it will resonate with a higher frequency. As the distance along the y-axis away from the centre increases, the resonant frequency decreases. The higher frequency from the centre of the element will not penetrate so far into tissue so that region will focus at shallow depths. The lower frequencies from a wider y-axis aperture will focus at greater depth. Overall, there is an improvement in slice thickness throughout the transducer range.

1.5-D Array

1.5-D array is also known as a 'one-and-a-half-D' array (Figure 4.15).

Each element is divided into three or more individual elements in the y-direction (elevation). They can then perform some degree of electronic focusing in the elevation plane to improve image quality.

2-D Array

2-D array is also known as a **matrix array.**

The matrix array has multiple elements in x- and y-direction (Figure 4.16). Ultrasound beams may then be controlled both laterally and in elevation to give the following advantages:

- The beam can be fully focused in both lateral and elevation planes giving better resolution and greatly reducing slice thickness artefacts (see Chapter 7).
- The beam can be steered in the x and y direction to give a 3-D data set from a volume of tissue.
- Collecting a 3-D volume data set over a period of time enables **4-D scanning** to be performed, e.g. to show complete information of heart valve motion across the cardiac cycle.

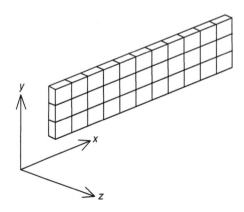

FIGURE 4.15   Arrangement of the elements of a so-called 'one-and-a-half-D' array.

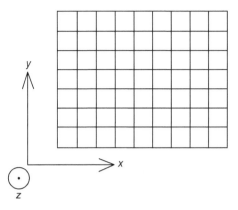

FIGURE 4.16   A 2-D array.

A matrix array will typically have 2000–50 000 elements. 3D data acquisition presents significant hardware demands in addressing a large number of elements and digitising and storing the information received. CMUT transducers lend themselves well to the construction of 2D transducers as they can be fabricated using microelectronic techniques. Sophisticated beamforming methods that use a reduced number of elements to achieve the same results as using a full array are being developed, sometimes called **sparse arrays** [4]. One example of such a technique is known as a row-column array [5, 6]. In terms of firing, the elements are arranged as a set of rows and columns as shown in Figure 4.17. By firing pulses from these two orthogonal

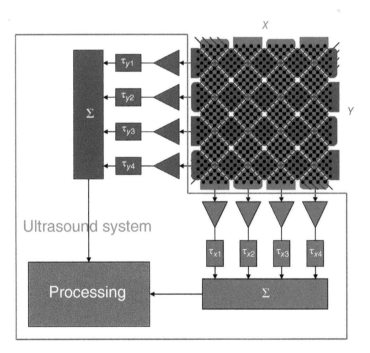

FIGURE 4.17   A 2D row-column array showing the set of rows and set of columns firing as two groups of elements. *Source:* Savoia et al. [5]/with permission of IEEE.

array sets in phase then in a second pulse firing the two arrays in opposite phase to one another, then subtracting the two pulses, 3D images can be formed that effectively use $2N$ elements in a probe instead of N^2, where N is the number of elements in a row.

REFERENCES

1. Ergun, S., Zhuang, X., Huang, Y. et al. (2005). Capacitive micromachined ultrasonic transducer technology for medical ultrasound imaging. *Proceedings of SPIE - The International Society for Optical Engineering* 5750: 58–68.
2. Cianci, E., Foglietti, V., Minotti, A. et al. (2006). Fabrication techniques in micromachined capacitive ultrasonic transducers and their applications. *MEMS/NEMS 2006*: 353–382.
3. Source: Binarysequence (2020). Licenced under CC Attribution-Share Alike 4.0 International. https://upload.wikimedia.org/wikipedia/commons/8/8d/Curved_Array_Ultrasound_Sensor_Construction.jpg (accessed August 2022).
4. Provost, J., Papadacci, C., Arango, J.E. et al. (2014). 3D ultrafast ultrasound imaging in vivo. *Physics in Medicine and Biology* 59: L1–L13. https://doi.org/10.1088/0031-9155/59/19/L1.
5. Savoia, A., Bavaro, V., Caliano, G. et al. (2007). P2B-4 crisscross 2D cMUT array: beamforming strategy and synthetic 3D imaging results. *IEEE Ultrasonics Symposium Proceedings* 2007: 1514–1517. https://doi.org/10.1109/ULTSYM.2007.381.
6. Jensen, J.A., Ommen, M.L., Øygard, S.H. et al. (2020). Three-dimensional super resolution imaging using a row-column array. *IEEE Transactions on Ultrasonics, Ferroelectrics, and Frequency Control* 67: 538–546. https://doi.org/10.1109/TUFFC.2019.2948563.

CHAPTER 5

Image Formation

The **pulse-echo principle** is the basis of nearly all diagnostic ultrasound techniques (Figure 5.1). A pulse of ultrasound is transmitted into the body and echoes arrive back at the transducer from ever greater depths with time after the pulse is transmitted. To get information from the target, the pulse has to travel twice the target distance d, i.e. out to, and back from the target, travelling at the speed of sound c in the medium, e.g. tissue.

$$\text{Echo arrival time} \quad \Delta t = \frac{2d}{c}$$

For soft tissue, $c = 1540\,\text{ms}^{-1}$ then, $\Delta t = 13\,\mu\text{s/cm}^{-1}$ 'go and return'

e.g. if a target is 6 cm deep, then echoes arrive back at the transducer $13 \times 6 = 78\,\mu\text{s}$ after transmit.

Ultrasound systems normally assume a speed of sound of $1540\,\text{ms}^{-1}$ in order to assign a depth to the origin of the echoes displayed in the image. This is known as the **system velocity**. It is the average speed of sound in soft tissue, which in reality varies by $\pm 5\%$ between different tissue types (Table 1.1). For some tissues, where the speed of sound is uniform but different to $1540\,\text{ms}^{-1}$, such as within the orbit of the eye, some equipment will allow the user to alter the system velocity in order to make accurate measurements.

Ultrasound Technology for Clinical Practitioners, First Edition. Crispian Oates.
© 2023 John Wiley & Sons Ltd. Published 2023 by John Wiley & Sons Ltd.

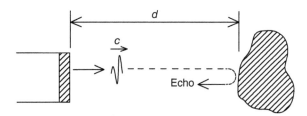

FIGURE 5.1   The pulse echo principle.

IMAGE MODES

The names used for the different types of greyscale image display come from the development of radar where the same types of display were first developed.

A-Mode (Amplitude Mode)

An 'oscilloscope' type of display showing the amplitude of the echo signal from different depths (Figure 5.2).

This type of display is not used much today but may be used in ophthalmology to calculate the power of the eye where the cornea, lens, and vitreous humour all have different speeds of sound. There are strong interfaces with echo free regions in between (Figure 5.3).

If the A-mode information is displayed using a bright spot whose brightness depends on the amplitude of the signal, then we produce the equivalent of one scan line on an image display.

M-Mode (Motion Mode)

M-mode allows moving targets to be displayed so their movement can be appreciated by the sonographer. Repeated pulses are fired along one single scan line and returning echoes are displayed as a series of A-mode bright-spot scan lines laid next to each other so as to show depth versus time. This gives an M-mode display where the change in position of returning echoes from along the scan line can be seen (Figure 5.4).

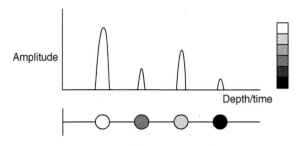

FIGURE 5.2   An A-mode scan line displayed as an oscilloscope display of amplitude and a series of spots whose grey level shows the signal amplitude.

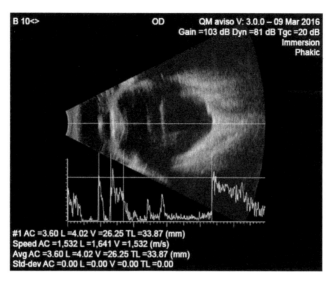

FIGURE 5.3 A-mode of the eye with B-mode image for comparison. The image shows a dense cataract and axial high-myopia. Note the separate speed of sound values used for anterior chamber AC, lens L, and vitreous V indicated on the second line of results. *Source:* Silverman [1]/Originally published by and used with permission from Dove Medical Press Ltd.

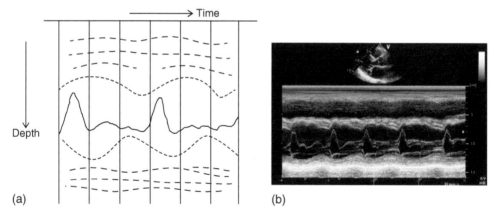

FIGURE 5.4 (a) Principle of an M-mode scan with (b) an M-mode across the left ventricle showing tips of mitral valve motion. *Source:* Courtesy: Chris Eggett.

- Stationary reflectors appear at constant depth with time.
- Moving reflectors show their change in depth from the transducer with time.

The scan line used to form the M-mode display is indicated on the image display and can be positioned to give information from the target of interest.

M-mode is primarily used in cardiac work to examine the movement of the heart wall and cardiac valves.

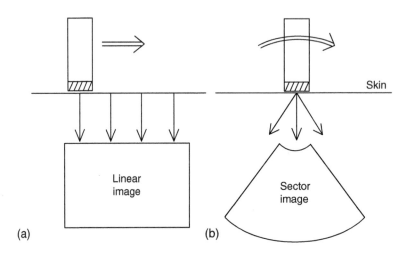

FIGURE 5.5   The scan pattern of (a) a linear image and (b) a sector image.

B-Mode or Greyscale Image (Brightness Mode)

This is the standard imaging mode used for most ultrasound investigations.

There are two basic image formats: **linear** and **sector** formats. These are the equivalent of moving the probe along the skin to gather an image slice or rocking the probe about a fixed point on the skin to sweep through a sector (Figure 5.5).

Each format has areas of advantageous use, for example,

- **Linear** – For vascular work, to look at lengths of straight vessels, where the angle of the probe surface to the vessel is maintained.
- **Sector** – Small footprint on the skin, but the image shape allows an extended organ to be seen, e.g. imaging the heart with access between the ribs or to image the extended length of a kidney.

The plane of the image through the body is known as the **image plane** or **scan plane**.

LINEAR IMAGE FORMATION

Linear Array

A **linear array** is an ultrasound probe formed with a straight line of closely spaced transducer elements (typical number, 160 elements). In order to achieve an aperture of sufficient width to form a narrow beam, several adjacent elements are connected electronically and fired together as one transducer (e.g. 8 elements). When the pulse-echo sequence from that group is complete, the element at one end of the group is dropped off and the next adjacent element at the other end is added to form a new

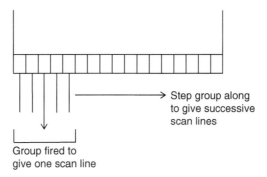

Step group along
to give successive
scan lines

Group fired to
give one scan line

FIGURE 5.6   The principle of stepping a group of active elements across a linear array.

group (Figure 5.6). The active aperture has then effectively stepped along the array by one element from the previous group. This is then fired with a pulse-echo sequence producing a beam that has moved across the front of the probe by a distance of one element from the previous beam.

By continuing this process right across the array, a sequence of beams is produced that sweeps across the width of the array. These beams obtain echo data across the whole scan plane to form an image that is a slice through the tissue being scanned.

The whole scan plane can be scanned in the time it takes, say, 120 beams to send out their pulse and obtain echoes from the depth of interest, e.g. $13\,\mu s \times 10\,cm \times 120 = 15.6\,ms$ for a depth of 10 cm.

Frame Rate

Frame rate is the number of times the complete **field of view** is scanned per second. In the example just given, this would produce a frame rate of 64 frames per second.

$$\text{i.e. Frame rate} = \frac{1}{15.6\,\text{ms}} = 64\,\text{fps}$$

This is fast enough to appreciate movement in **real time**. In practice, the frame rate would be somewhat slower than this to allow for echoes from the previous pulse to die away and can be changed by the image acquisition techniques used as discussed below and in Chapter 7.

Curvilinear Array

A **curvilinear array** uses the same method of stepping along a group of elements to form an image as the linear array, but with the array formed into a curve to give a sector shaped display, giving, for example, a wide field of view in the abdomen (Figure 5.7).

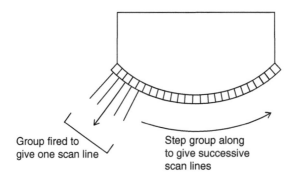

FIGURE 5.7 The principle of stepping a group of active elements across a curvilinear array.

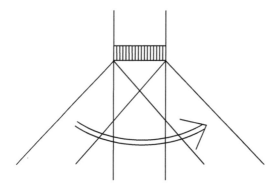

FIGURE 5.8 The scan sector produced by beam steering in a phased array.

Phased Array

A **phased array** uses beam steering to form a sector shaped image using the whole array as the active aperture for every pulse (Figure 5.8). By forming the probe with very closely spaced elements, less than a wavelength apart, the beam can be steered through a large angle without incurring the problem of grating lobes (see Chapter 3). This enables the probe to have a small footprint on the skin but with the image plane spreading out inside the body to image an extended organ. For example, imaging the heart from between the ribs.

Extended Field of View

Extended field of view uses beam steering to increase the field of view of a linear array at the extremities of the image (Figure 5.9).

Panoramic View

Another way in which the field of view may be extended is to store a sequence of images as the probe is physically moved across the surface of the skin along the line of

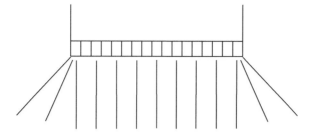

FIGURE 5.9   Using beam steering to extend the field of view of a linear array.

the scan plane (Figure 5.10). The images are then electronically matched and blended in a similar manner to the panoramic view setting on a camera. By this means an image may be formed, say, of a large cyst or a blood vessel that is bigger or longer than the field of view of a single image (Figure 5.11). The probe must be moved smoothly across the skin by the operator, and the calibration of lateral distance on the image will be poorer than for an image formed when the probe is stationary. The image is built up over a period of time, as the probe is manually moved, so the image produced is not a real-time live image.

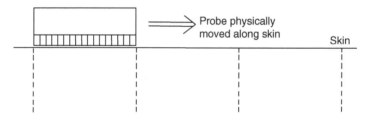

FIGURE 5.10   The principle of producing an extended field of view by physically moving the probe on the skin.

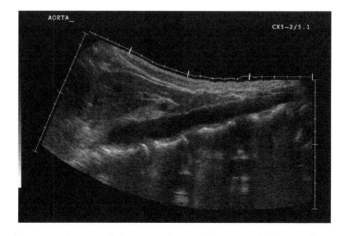

FIGURE 5.11   A panoramic scan of the aorta. *Source:* Courtesy of Simon Elliott.

3D IMAGING

Three-dimensional (3D) imaging of a block of tissue may be obtained in one of three ways.

Hand Scanning

In the same manner that an extended field of view may be obtained in the scan plane, by physically moving the probe across the skin in the scan plane, a 3D image may be obtained by manually sweeping the probe across the skin in a direction perpendicular to the scan plane (Figure 5.12). These images are then able to give a 3D data set through a block of tissue. In this technique, the calibration of distance in the out-of-scan plane axis (y-axis) cannot be relied upon. As with the panoramic view, the image produced is not a real-time live image.

Mechanical 3D Probe

A curvilinear array is housed in a probe where it is made to oscillate by a motor in the out-of-scan-plane direction as shown in Figure 5.12. A 3D data set is obtained. This may be used for measurement as the direction the probe is pointing is known at all times, so true distances may be calculated. Data acquisition is fast enough to give a real-time display but with a relatively low frame rate [2].

2D Matrix Array

A two-dimensional (2D) electronic array of elements may be electronically scanned in the two perpendicular planes, using beam steering, to produce a 3D data set in real time (Figure 5.13).

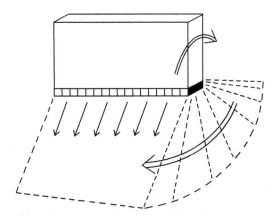

FIGURE 5.12 Scanning a 3D volume of tissue by physically rocking the probe in the elevation plane.

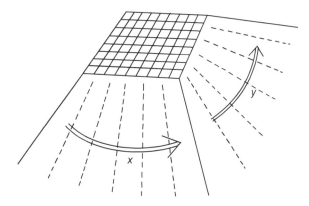

FIGURE 5.13   Producing a volume scan by steering in the *x*- and *y*-plane with a 2D array.

3D Image Display

Having acquired a 3D data set, images may be displayed from any of the orthogonal or other planes in the volume data set (Figure 5.14). For example, a set of slices through an organ (Figure 5.15). This includes scanning in a plane that is orthogonal to the direction of propagation of the ultrasound beam at constant depth from the transducer. Such a scan is in the coronal plane and has been called a **C-scan**.

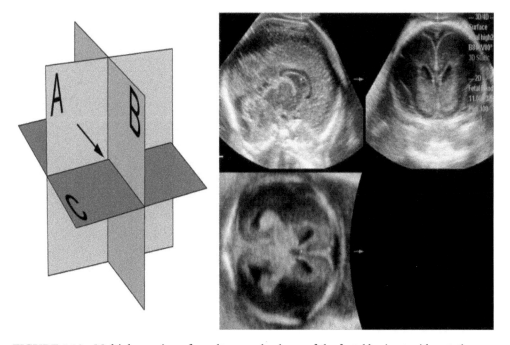

FIGURE 5.14   Multiplanar view of an ultrasound volume of the foetal brain at mid-gestation obtained with a transvaginal approach from the sagittal plane B. *Source:* Pilu et al. [3]/with permission of John Wiley & Sons.

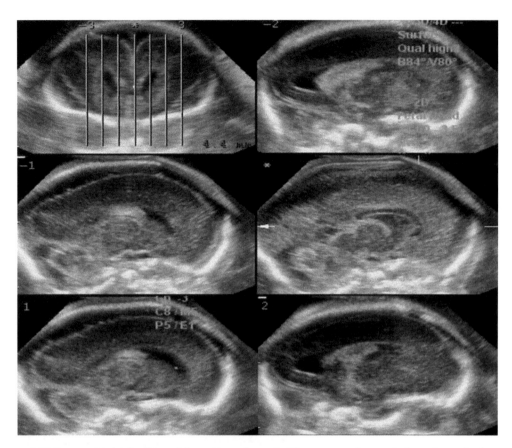

FIGURE 5.15   Multiple slices of the volume shown in Figure 5.15 orientated along the sagittal plane. *Source:* Pilu et al. [3]/with permission of John Wiley & Sons.

CINE LOOP

Cine loop is the name given to the storage of a sequence of images that may be played back for review. Where a 3D data set is recorded and stored to form a stored time sequence, the term **4D scanning** is used, with the fourth dimension being time.

ENDOPROBES

An endoprobe is one that can be introduced inside the body via a natural body cavity or via a minimally invasive surgical wound, e.g. endovascular, oesophageal, rectal, and transvaginal probe (Figure 5.16).

These probes may have a linear array, a tight curvilinear format or may be formed from a set of elements around the probe to produce a **radial image.** In a radial image, the ultrasound beam is swept round like the light from a lighthouse (Figure 5.17). This produces an image plane like a radar view with the

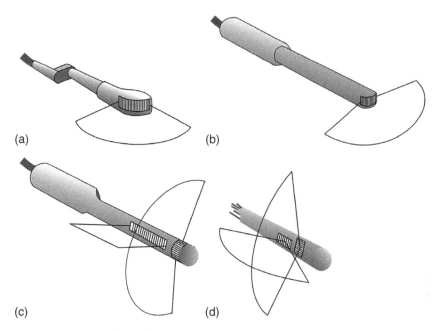

FIGURE 5.16 Examples of scan formats on endoprobes. Panels (a) and (b) have tight curvilinear arrays as might be found in transvaginal probes, (c) has a linear array and a radial array as might be found on a transrectal probe to examine the prostate, and (d) has two orthogonally arranged phased arrays as used in a transoesophageal probe for echocardiography. *Source:* Whittingham [4]/ with permission of Elsevier.

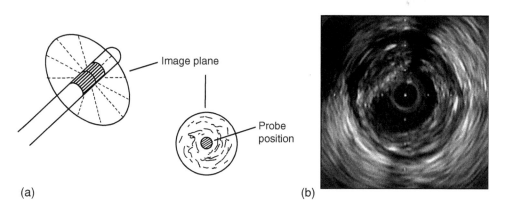

FIGURE 5.17 (a) Illustration of a radial image produced by an endoprobe with radial elements. (b) An example of a radial image showing plaque in a coronary artery. *Source:* (b) Nissen and Yock [5]/with permission of Wolters Kluwer Health, Inc.

probe in the body cavity at the centre of the field of view. An example of this form of imaging is intravascular ultrasound, which is used to examine diseased arteries. By using a pull-back technique, a 3D data set of the surrounding tissue may be obtained.

CHOOSING A PROBE

The choice of probe for a particular application will depend on a number of factors. The imaging target will be the prime factor and will usually dictate whether a linear, sector, or endoprobe format is used. There may be constraints on the footprint of the probe for access into awkward angles of approach to a target, e.g. the use of a phased array for cardiac work accessing between the ribs, or a tightly curved curvilinear array for access to a neonate fontanelle. For peripheral vascular work or small parts scanning a linear array may be most appropriate.

The second factor is **image resolution**. Here, consideration must be given to resolution versus penetration to the target of interest. *The probe with the highest frequency that allows the target of interest to be adequately imaged should be chosen.*

Where 3D data are required a probe with that ability must be used.

Resolution will also be affected by the focusing used and the frequency used within the probe bandwidth.

FOCUSING

We saw in Chapter 3 that focusing the ultrasound beam gives an improvement in beamwidth and hence resolution at the focal point. Outside the focal zone, the resolution is poorer as the beam spreads out. Focusing to several depths is called **multizone focusing**. This can be achieved using electronic focusing and is used to improve the resolution at all depths in the image.

Successive pulses do not have to have the same shape as each other. The focal depth between each pulse can be changed. Neither do the transmit beam and receive beam have to have the same shape. Using this fact, resolution can be improved across the whole image by choosing appropriate focal zones for each pulse.

Transmit Focus

This is the focus used for the transmitted pulse.

Once a pulse has left the transducer, no more processing can be done to it. Therefore, each transmitted pulse can only be focused to a single depth. In other words, the focal depth is static for each pulse echo sequence. To focus at a different depth along that scan line requires another transmitted pulse focused to the new depth. In order to focus to three different depths in transmit, it will require three transmit pulses (Figure 5.18). For each transmitted focal zone, echoes are only detected for the depth at which the beam is focused. Doing this for each scan line in the image will reduce the frame rate by a factor of 3.

Receive Focus

This is the focus used for the received echoes.

On receiving, we can rapidly switch beam shapes for echoes coming from different depths. For example, we can focus to 2 cm deep when echoes are coming

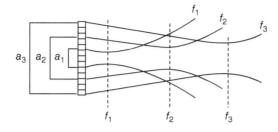

FIGURE 5.18 Using different apertures and delays to produce a set of focal zones at different depths.

from 2 cm deep and then switch to a receive focus of 4 cm deep when echoes are arriving from 4 cm deep, etc. In other words, the focal zone can be set for each successive depth for which echoes are returning. By switching in, say, 10 focal zones in quick succession we can construct a very narrow receive beam (Figure 5.19). This is known as **dynamic receive focusing** as the focal zone changes within a single pulse-echo sequence.

All machines will automatically use dynamic multi-zone focusing for the receive beam so as to give the best resolution possible.

For the transmit focus the user must select the depth of the focal point. This should be at the target of interest so that the target is imaged with the finest resolution. The user can choose to use more than one focal zone on transmit with the penalty of reduced frame rate.

NOTES

The overall resolution of the image will result from the combination of the transmit and receive beam shapes multiplied together to produce the image.

There are techniques whereby **dynamic transmit focusing** can be realised, extending the focal zone within the image. The most effective way of achieving this is through synthetic aperture techniques described in Chapter 13.

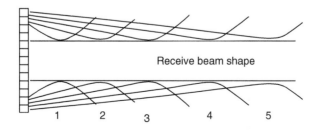

FIGURE 5.19 The narrow beam produced by using a number of focal zones in receive.

INCREASING FRAME RATE

Conventional B-mode imaging frame rates are limited by the need to wait for the go-and-return time for pulses for each beam in the image. Frame rate may be increased by using multiple beams to increase the rate of acquisition of image data.

One way to achieve this is to transmit a wider beam than normal and receive the echoes through 2 or 4 beam formers to create 2 or 4 receive beams for each transmit beam. This will be at some cost to lateral resolution but will increase frame rate by a factor of 2 or 4. Alternatively, multiple beams can be transmitted and detected simultaneously at positions where they do not interfere with one another. Using such techniques, real data frame rates of 60 fps can be achieved.

Synthetic Aperture Imaging

Synthetic aperture imaging is a sophisticated technique whereby an ideal ultrasound source is mimicked by firing elements in an array so as to produce a beam that would have come from the ideal source. By creating such sources and combining the received echoes appropriately, transmit and receive focuses can be synthesised for every point in the field of view. This can be achieved at very high speed with greatly improved frame rates. It also improves contrast and resolution throughout the image as every point is focused on transmit and receive.

Chapter 13 describes in detail how synthetic aperture imaging works.

> **NOTE**
>
> Some scanners using these techniques do not require the user to set the transmit focal depth.

USER CONTROL

Resolution/Penetration

Wideband transducers allow a range of frequencies to be transmitted and detected. The user has some control over the frequencies used to form the image. This is often presented in the form of a control for 'Resolution/Penetration'. This may be adjusted by the user to obtain the best image. It is trading frequency for penetration within a single probe.

Another technique to improve resolution by using higher frequencies is **harmonic imaging**.

Background Theory to Understanding Harmonic Imaging

1. Superposition of Waves

If two waves coincide then the amplitude at each point is the sum of those two wave's amplitudes at that point, e.g. a peak plus a peak gives a bigger peak, a peak plus a trough cancel each other out as seen in Chapter 2.

Figure 5.20 shows the superposition of two waves where one wave has exactly twice the frequency of the other.

NOTE

If the phase of $2f_0$ relative to f_0 changes then the shape of the resultant wave will also change.

2. Harmonics

If the frequency of a given sine wave f_0 is increased by an integer multiple, i.e. ×2, ×3, ×4, etc., we generate a set of **harmonic frequencies**. In music, a doubling of frequency gives a note one octave higher (Figure 5.21).

f_0 is called the **fundamental** or **first harmonic.**
$2f_0$ is called the **second harmonic.**
$3f_0$ is called the third harmonic.
$4f_0$ is called the fourth harmonic.

Harmonics can be heard on a stringed instrument, or seen by shaking a skipping rope, by forcing the vibrations to divide the string into half, third, etc. (Figure 5.22)

The French mathematician Fourier showed that any repeating waveform can be shown to be made up of the superposition of a fundamental + harmonics of higher multiples of f_0.

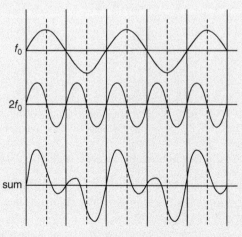

FIGURE 5.20 The superposition of a wave of frequency f_0 and a frequency of $2f_0$.

(continued)

(*continued*)

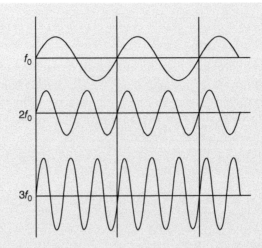

FIGURE 5.21   The fundamental and first two harmonics of a wave of frequency f_0.

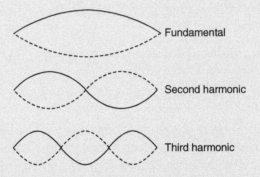

FIGURE 5.22   The first three harmonics as seen on a vibrating string.

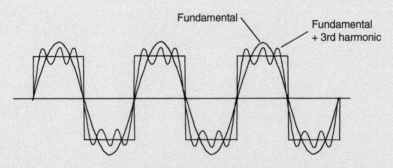

FIGURE 5.23   Diagram showing how a square wave is built up of odd harmonics.

The higher harmonics usually have smaller amplitudes.

Example: A square wave (repeating) is made up of the fundamental + odd harmonics (3,5,7,9. . ..) (Figure 5.23).

The more harmonics that are added, the closer we get to a square wave. Adding *all* the odd harmonics gives a pure square wave

Here, we can see that it is the *higher harmonics* that turn a sine wave into a square wave. It is generally true that if a waveform has sharply rising/falling edges, then there must be higher harmonics in its makeup.

SUMMARY

- Any repeating signal can be broken down into a set of harmonic frequencies superimposed on one another – the amplitude and relative phase of each harmonic component must be right (like a cooking recipe – too much salt does not work).
- f_0 is the fundamental frequency and first harmonic. Harmonics are multiples of the fundamental.
- If a waveform has sharp edges, then it must include higher 'faster moving' frequencies in its composition.

ULTRASOUND HARMONICS

Up to now, we have assumed that a pulse travels through tissue without any change to its shape, and it behaves in what is called a '**linear**' way. That is, if you increase the intensity the wave has a bigger amplitude but still has the same shape. In reality, as a wave or pulse travels through tissue it becomes distorted due to **non-linear propagation.** This occurs because the speed of sound at peak pressure is slightly higher than the speed of sound at minimum pressure. The result is that the peaks move faster than the troughs and tend to catch them up (Figure 5.24). This is like the peaks of sea waves getting steeper and breaking on a beach as the sea gets shallower. The peak catches up with the next trough.

> Question: What happens to the spectrum (frequency content) of the wave as it changes shape?
> Answer: Second (+other) harmonics are generated. As we saw in the theory above, sharper edges imply that higher frequencies must be present.

Energy goes from the fundamental into the second harmonic, i.e. non-linear propagation (Figure 5.25). The Spectra shown in Figure 1.18 show the second harmonic in a B-mode and a Doppler pulse due to non-linear distortion in the pulses.

We know that higher frequencies give better resolution, so if we can use the second harmonic, so generated, to form our image, the resolution of the image will be

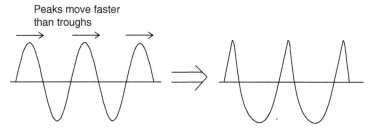

FIGURE 5.24   The effect of non-linear propagation causes the peaks to get sharper and the troughs to become wider.

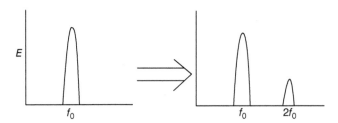

FIGURE 5.25   The energy spectrum of a wave showing non-linear behaviour. Energy is lost from the fundamental f_0 and goes into the generated second harmonic $2f_0$.

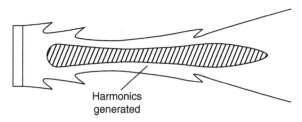

Harmonics
generated

FIGURE 5.26   Showing the area within the beam where intensities are great enough to generate harmonics. *Source:* Adapted from Whittingham [4].

improved. Using the second harmonic to form the image is called **harmonic imaging.** In addition to higher frequencies giving improved resolution, the 'beam' produced at the harmonic will be narrower. This is because harmonics are produced at higher amplitudes and the highest amplitudes are seen along the main axis of the beam, as shown in Figure 5.26. This 'harmonic beam' has no side lobes, so clutter is also reduced.

> ### Aside: Filters
>
> **High-Pass Filter** – A device that allows high frequencies above its **cut-off frequency** to pass, and blocks low frequencies.
>   **Low-Pass Filter** – A device that allows low frequencies below its cut-off frequency to pass and which blocks high frequencies.
>   These devices may be electronic devices or be implemented as a software algorithm to process a digitized signal.

Harmonic Imaging

There are two basic ways we can obtain the harmonic information to form an image.
    Using a **filter method** (Figure 5.27):
    A wideband transducer will enable both the fundamental frequency to be transmitted and the second harmonic to be detected. If a high-pass filter is used to cut out f_0 from the received signal, then the image can be formed from $2f_0$. Slightly longer pulses than for normal B-mode are used to narrow the pulse bandwidth and improve

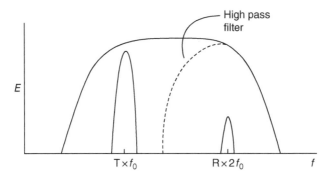

FIGURE 5.27   The use of a high pass filter to detect the second harmonic signal within the transducer bandwidth.

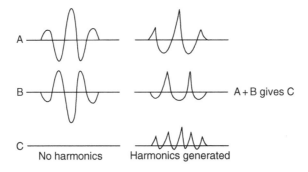

FIGURE 5.28   The pulse inversion method of detecting the harmonic signal. The echoes from two consecutive pulses A and B are added together to give the resultant signal C.

separation of first and second harmonic for filtering. This will slightly degrade the axial resolution. But overall there will be an improvement in resolution from using the higher frequency harmonic to form the image.

Using a **pulse inversion method** (Figure 5.28):

This is the more common implementation and uses normal length short pulses. It requires two transmit pulses of opposite phase to be transmitted in sequence. The echoes from the two pulse sequences are stored and added together. At the fundamental frequency where the pulse maintains its shape, the two echo trains cancel each other out. Where harmonics are generated, there will be a net signal that can be used to generate a harmonic image. Axial resolution is maintained at a cost of frame rate.

Pros and Cons of Harmonic Imaging

Pros

- Better resolution with a higher frequency and narrower beams.
- Removes 'clutter' from the image (due to reduction in sidelobes, reverberations and multiple echo paths)

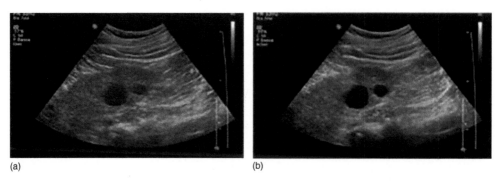

(a)   (b)

FIGURE 5.29   Image of renal angiomyolipomas showing the decluttering effect of harmonic imaging. Image (a) has no harmonic processing and image (b) has harmonics activated. *Source: Khandelwal et al. [6]/OAE Publishing Inc., CC BY 4.0.*

Cons

- Weaker echoes at $2f_0$ with higher attenuation so penetration is poorer.

In practice, the harmonic imaging mode on a scanner uses a combination of second harmonic and first harmonic to give an overall improvement to the image without losing penetration. Its effect is seen in Figure 5.29.

Harmonic Imaging is something that the user can choose to turn on or off.

CODED EXCITATION

Coded excitation, sometimes called **pulse compression**, is a technique used to improve sensitivity to weak echoes. Echoes from the very short pulses used for imaging are hard to detect in the presence of significant background noise and clutter. Very short pulses 'look like' noise spikes. One way to improve **signal-to-noise ratio (SNR)** would be to increase the transmitted power but that has safety implications. By imposing a 'code' on to the transmitted pulse, the echoes can be decoded and the true echo signal recovered from a noisy background.

Instead of a short high intensity pulse, a longer pulse with lower average intensity that carries a code is used. The longer pulse would give poorer resolution except that when it is decoded with a complimentary code, the result is a compressed output of the same length as a normal short pulse. This can give decoded pulses with a larger peak intensity than would have been obtainable from a short pulse within safe intensity limits. Hence, the term 'pulse compression' is used.

Examples of Coded Excitation

There are a number of ways a pulse can be encoded but to give an idea of what is involved the following examples are given:

Transmit two pulses which are shaped with a pseudo-random code A and a complementary code B. Their echoes are received. To decode, the echoes from each pulse are matched to a time reversed copy of the transmitted pulse in an cross-correlator as shown in Figure 5.30 (see Chapter 9). The correlated outputs are then summed together. The result is a strong signal for the encoded echoes whilst noise in the signal tends to cancel out.

The symbol ⊗ means cross-correlation.

The encoding may use changes in the phase of the signal to give a binary +1, −1 code, or the encoding may be produced by changing the transmit driving frequency during the course of the pulse.

In the example in Figure 5.31, the code is shown as changes in phase within the transmitted pulse and the code used only requires one transmission-decoding sequence.

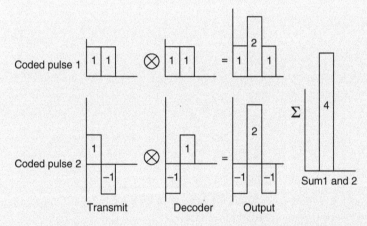

FIGURE 5.30   Diagram demonstrating the principle signal detection using coded excitation.

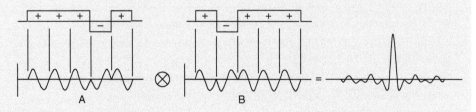

FIGURE 5.31   Coding of the signal using 180° phase shifts and the detected signal.

Chirp Pulse

A chirp pulse is a pulse whose frequency varies throughout the pulse such that the driving frequency is swept through a range of frequencies from the start to the end of the pulse (Figure 5.32). A matched filter in the receiver ensures that only echoes matching the chirp are detected enabling true echoes to be discriminated from background noise.

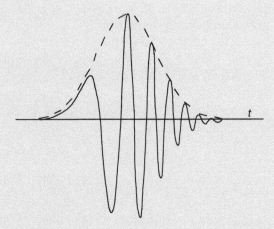

FIGURE 5.32   Example of a chirp pulse covering a range of frequencies as the frequency increases throughout the pulse.

REFERENCES

1. Silverman, R.H. (2016). Focused ultrasound in ophthalmology. *Clinical Ophthalmology* 10: 1865–1875. https://doi.org/10.2147/OPTH.S99535.

2. Huang, Q. and Zeng, Z. (2017). A review on real-time 3D ultrasound imaging technology. *BioMed Research International* 2017: Mar 26. https://doi.org/10.1155/2017/6027029.

3. Pilu, G., Ghi, T., Carletti, A. et al. (2007). Three-dimensional ultrasound examination of the fetal central nervous system. *Ultrasound in Obstetrics & Gynecology* 30: 233–245.

4. Whittingham, T.A. (2007). Medical diagnostic applications and sources. *Progress in Biophysics and Molecular Biology* 93: 84–110.

5. Nissen, S.E. and Yock, P. (2001). Novel pathophysiological insights and current clinical applications. *Circulation* 103: 604–616. https://doi.org/10.1161/01.CIR.103.4.604.

6. Khandelwal, N., Chowdhury, V., and Gupta, A.K. (2013). *Diagnostic Radiology: Recent Advances and Applied Physics in Imaging.* Jaypee Brothers Medical Publishers. ISBN 10: 9350904977.

CHAPTER 6

The B-Mode Scanner

Looking at the basic layout of an ultrasound B-mode imaging system gives an overview of the basic processes involved in producing an ultrasound image and shows where the main controls accessible to the user are enabling them to interact with the image forming process. We begin with the transmission side and then look at the receive side of a scanner.

TRANSMISSION SIDE OF A SCANNER

The transmission side of a B-mode scanner is shown in Figure 6.1.
Modern scanners can:

- Alter the shape of the transmit (Tx) pulse.
- Change the transmit power between individual elements.
- Change the timing of the pulses between elements and control the phase of the pulses at each element.
- Change the frequency of the pulse.
- Change the aperture used.

and it can do all of these using different values from one pulse to the next, i.e. no two pulses need be the same.

Ultrasound Technology for Clinical Practitioners, First Edition. Crispian Oates.
© 2023 John Wiley & Sons Ltd. Published 2023 by John Wiley & Sons Ltd.

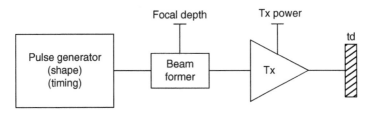

FIGURE 6.1   Block diagram of the transmit side of B-mode scanner.

This enables:

- Dynamic multi-zone focusing
- Beam steering
- Beam apodization to be used to suppress side lobes and reduce 'clutter' in the image.
- Control the phase of the pulse for harmonic imaging
- Simultaneous B-mode and Doppler imaging using different transmit frequencies
- Allow complex data acquisition, combining echoes from multiple apertures to achieve uniform focusing throughout the image and faster frame rates.
- Generate encoded wave trains – coded excitation – to improve signal to noise ratio of weak signals.

Beam Former

The **beam former** controls exactly which pulse – shape, timing, phase – is produced by each element in the transducer to give the required focal depth and beam steering for each transmitted ultrasound pulse.

PRF – Pulse Repetition Frequency

This is typically 1000s of pulses per second. (recall: the go and return time to 10 cm depth is 130 μs which is equivalent to a PRF of 7690 pulses per second)

USER CONTROLS

Transmit Power

Transmit power needs to be adjusted to minimise risk of injury from energy deposited in the tissue whilst giving adequate penetration to image the target of interest.

Depth of Focus

The depth of focus needs to be set at or just below the target of interest. More than one focal zone may be chosen to extend the focused field of view at the expense of frame rate.

RECEIVE SIDE OF A SCANNER (Rx)

The receive side of a B-mode scanner is shown in Figure 6.2.

Beam Former

In receive the beam former sums the received echoes from each of the transducer elements with appropriate time delays so as to form the 'receive beam' with the correct focuses and steering.

Gain

As can be seen from Figure 6.3, the size of echoes has a very large range. In general, the echo signal from within tissue, carrying the information we want to see, is very weak and the signal must be amplified. **Gain** is another word for amplification.

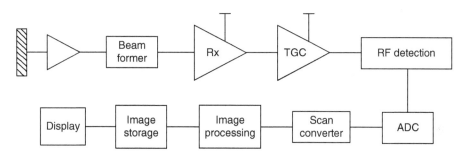

FIGURE 6.2   Block diagram of receive side of B-mode scanner.

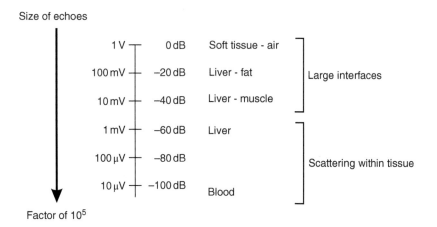

FIGURE 6.3   Size of echoes received from different tissue interfaces (note: the voltage is signal amplitude, decibel values are for signal power).

User Control – Gain

- The overall signal gain may be controlled by the user. This will lift or depress the overall brightness of the image on the display.
- If the gain is set too high, mid to high level echoes will be pushed towards peak white on the display and information may be lost. The image will also become noisy as low-level noise is over amplified.
- If the gain is set too low information in lower level echoes may be lost due to under amplification.
- The gain should be adjusted so that the peak signal in the image just shows as peak white whilst the lower level echoes that give the clinical information are all distinguishable on the grey scale.

Time Gain Compensation

We saw in Chapter 2 that attenuation causes returning echoes strength to decrease with distance from the transducer. The received signal therefore needs to be amplified more and more to compensate for the loss in signal strength. This process is known as **Time Gain Compensation (TGC)** – also known as **'swept gain'** or **Depth Gain Control (DGC)**.

Without correction, the image would be bright near the surface and get gradually dimmer for greater depths in tissue as returning echoes get weaker due to attenuation from absorption, scattering, and beam divergence. To be clinically useful, the image from a large organ, such as the liver, should have a uniform brightness at all depths so that no clinically significant changes in appearance are missed. For each transmit pulse, **G**ain that increases with **T**ime of echo arrival is applied to **C**ompensate for attenuation of the pulse, i.e. **Time Gain Compensation** (or **TGC**). If the correct amount of gain is applied, the signal strength from a uniform medium will remain constant for all depths over the usable depth of penetration of the probe as shown in Figure 6.4.

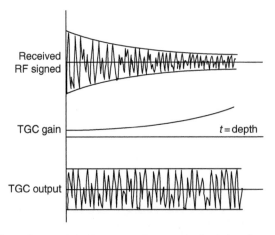

FIGURE 6.4   The effect of Time Gain Compensation on received signal amplitude.

SUMMARY

- Due to attenuation, the received echo signal will reduce in amplitude for echoes arising from further away from the transducer. On average, the level falls off exponentially with depth of target.
- The TGC gain curve is applied to the returning echoes from each transmit pulse.
- Applying TGC gives a uniform signal over the useable range of echoes. At large depth no detectable echoes return and only electrical noise in the system is amplified. The depth at which this happens defines the depth of penetration for a particular transducer.
- Most machines will apply a basic TGC to returning signals by default.
- Fine adjustment must be controlled by the user via a set of sliders or other controls that adjust the gain at various depths in the image so as to give a uniformly bright image over the useable penetration of the ultrasound probe.
- Some machines use sophisticated algorithms to improve image uniformity automatically based on soft tissue speckle statistics.

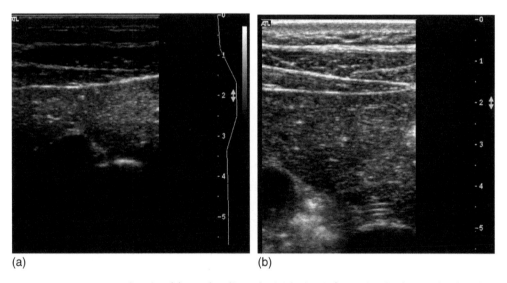

(a)   (b)

FIGURE 6.5   Images showing (a) poorly adjusted TGC losing information in the proximal and distal zones and (b) the same scan correctly adjusted.

An example of how to adjust TGC correctly is given in Figure 6.5.

Analogue to Digital Converter – ADC

The ADC converts the analogue signal from the echoes into a digital signal where each point in the signal is represented by number that can be stored in a digital memory.

How Analogue to Digital Conversion Works

Example of a time varying signal (Figure 6.6):

Following digitisation the signal is described by the number sequence 3, 5, 7, 8, 7, 6, 7, 8, 9, 10, 8, 5, 1.

In each time interval T_1, T_2, T_3, ... the analogue value of the amplitude of the echo signal is converted to the nearest integer value and represented by that number. The complete echo train is then stored as a sequence of numbers (in binary format).

> **NOTE**
>
> There is some loss of information as only integer values are stored, e.g. there is no value for 5.7, it will be converted to 5.

The quality of the data therefore depends on A–D rate (how many times per second a value is calculated) and the number of binary bits the signal is digitised to. The amplitude of an echo signal from a 10 MHz transducer will change at least 20 M times per second (going positive then negative).

Digitisation must therefore be carried out at least 20 M times per second in order to give a faithful representation of the signal in digital form. Many scanners will digitise the raw echo signal at 40 MHz.

Using binary representation of the digitized number each binary digit can be 0 or 1. The number of bits used to represent each digitization point of the signal will determine how the greyscale image will appear to the viewer.

Number of bits to convert to:

1 bit	0 and 1	= black or white only
2 bits	00,01,10,11	= black, white and 2 grey levels in between
8 bits	2^8	= 256 grey levels
12 bits	2^{12}	= 4096 grey levels – the human eye cannot distinguish this number of levels but it results in a much smoother looking image.

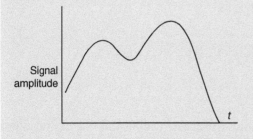 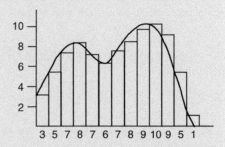

FIGURE 6.6 Digitisation of an analogue signal.

Advantages of Digitising

- No further risk of introducing noise or distortion into the signal
- Powerful computing can be done = image processing
- Easy storage in memory

Echo Detection

This is the process whereby the raw echo signals are detected and turned into image data. The raw signal is called the **RF signal** (for radio frequency). It is a time varying signal whose phase and amplitude are constantly changing. The B-mode image displays the **amplitude** of the echo arising from each point in the field of view.

NOTES

- The amplitude of the analogue signal may be directly detected and used to form the B-mode image.
- Modern scanners often digitize the raw RF signal of the returning echoes, preserving phase information to allow advanced signal processing techniques to be used. The digital amplitude of the processed signal is then used to form the B-mode image.

Detection of the Analogue RF Signal

The echo train from a transmitted pulse arriving at the transducer is a complex signal with varying amplitude from many reflecting targets in the body (Figure 6.7). The signal itself typically varies at the transmit frequency of the transducer. The size of echoes arising from targets in the tissue will change the **amplitude** of the echo signal as seen in the outline or **envelope** of the echo signal (Figure 6.8).

Consider One Pulse and One Reflecting Target

The echo signal is **rectified** (made all positive) and passed through a low pass filter. This cuts out all the high frequency MHz components leaving the envelope of the shape of the echo pulse. Rectification and low pass filtering together is called **envelope detection** and together they produce a **demodulated echo**.

FIGURE 6.7   Example of the RF echo signal received at a transducer.

(continued)

(*continued*)

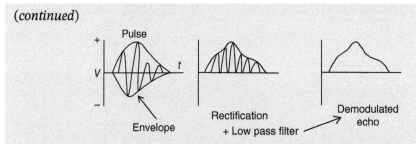

FIGURE 6.8   The principle of envelope detection with rectification to form a video pulse.

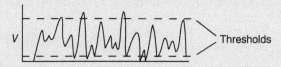

FIGURE 6.9   Thresholds applied to limit signal levels.

The signal may be scaled and cut-off thresholds may be applied to remove low level noise and limit the signal to a useable range as shown in Figure 6.9.

DYNAMIC RANGE AND TRANSFER FUNCTION (GREYSCALE MAPPING)

Dynamic Range is the difference in signal strength between the strongest signal and the weakest signal detected or displayed.

Contrast is the difference in grey level between adjacent targets in the image. For example, the appearance of an image as black and white between adjacent areas is a high contrast image versus subtle changes in grey level which would be a low contrast image.

At the low end of the range of signal strength will be 'noise'. Some of this noise will be from extremely low-level echoes but it will also include electrical noise from within the scanner. The scanner may include a **noise rejection** control to remove this very low-level signal that does not contribute to the useful information in the signal. This is essentially a filter that blocks all signals below a chosen amplitude level.

The dynamic range can be controlled, limiting the size of signals to a narrow or wider output range. This process of limiting the signal to a given range is called **signal compression.**

NOTE

The **receive gain** used also interacts with the dynamic range by lifting or lowering the echo levels within the dynamic range. For example, if the gain is set very high the noise at the lower end of the echo signal range will extend over a greater proportion of the dynamic range and less bright echoes will get pushed to toward the top white level of the range. The overall appearance of the image will be brighter but the image will be more noisy and important information may be lost.

TABLE 6.1   Showing How the Number of Bits Converts to Dynamic Range

Bits	Value Range	Dynamic Range (dB)
4	0–15	24
8	0–255	48
10	0–1023	60
12	0–4095	72

Using the decibel scale (see Appendix B) for a digitized signal (binary 0,1 bits), we can see how the dynamic range relates to the number of bits a signal is digitized to (Table 6.1).

As shown in Appendix B, doubling of amplitude $= +6\,dB$

so $4\,bits = 2 \times 2 \times 2 \times 2 = 16\,levels$ or $4 \times 6\,dB = 24\,dB$ etc.

Dynamic Range

As shown in Figure 6.10, a wide dynamic range with a large number of grey levels will give a 'smooth' appearance to the image. A narrow dynamic range with more signal compression will give a more 'contrasty' image with a more black-and-white appearance.

- Echoes from the body have a dynamic range of 100 dB
- The human eye has a dynamic range at one time of ~30 dB (i.e. without adjustment over time to dark/bright environment etc.)
- A good quality flat screen display monitor has a dynamic range of 60 dB

Transfer Function or Grey Scale Mapping

We therefore need to use **signal compression** to 'fit' the large dynamic range of the echoes into the dynamic range of the display monitor. The relationship between the input and output signal is called the **Transfer Function** or **Grey Scale Mapping**.

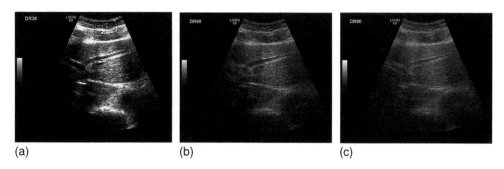

(a)   (b)   (c)

FIGURE 6.10   Showing the effect of increasing dynamic range (a) 36 dB, (b) 69 dB, (c) 90 dB.

An example of such a transfer function would be to scale the dynamic range of the echo signal to fit the range of the display monitor using a linear scale as shown in Figure 6.11.

Such a linear grey mapping gives equal emphasis to every change in grey level in the image. Although this may seem a good idea, it is not usually the best choice for diagnostic scanning. Clinically the most useful echoes are those in the mid-range of amplitude. Very strong reflectors such as bone or gas filled structures appear as bright white in the image and are clearly seen. At the bottom end, the signal becomes noisy and it is using valuable grey scale span to emphasise these poor echoes. Therefore, a mapping that emphasises the mid-range is a good way of distributing the diagnostically most useful echoes across the available range of grey levels. With reference to Figure 6.12, this leads to an 'S' shaped transfer function (a). Such a mapping de-emphasises the very low-level echoes and the brightest echoes and emphasises the mid-range echoes. Another choice would be to emphasise the mid-low level echoes (b) where subtle changes in tissue texture may be seen, such as when looking for liver metastases.

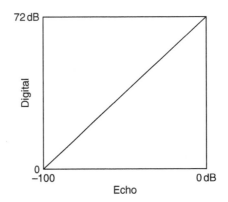

FIGURE 6.11   Signal compression using a linear transfer function.

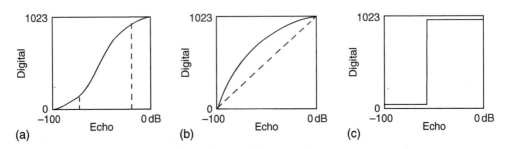

FIGURE 6.12   Examples of transfer functions. (a) An "S" shaped curve emphasising the mid-range echoes, (b) a curve emphasising low level echoes and (c) a step curve giving a binary black and white image.

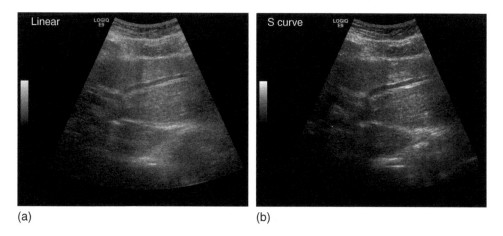

(a)   (b)

FIGURE 6.13   Illustrating a change in transfer function from a linear curve (a) to a S-shaped curve (b).

The choice of mapping can dramatically alter the appearance of the image. Even if a wide dynamic range is used, the choice of a steep mapping can produce a very 'contrasty' image. For example, in the extreme case, the step mapping shown in (c) would give a black and white binary image. Every signal to the left of the vertical line would show as black and every signal to the right would show as white. The level of that black-white cut-off depends on where within the dynamic range the cut-off is placed. In practice such a binary cut-off would never be used, but to think about it illustrates the effect the mapping has relative to the dynamic range used. Examples of linear and S-shaped transfer functions are shown in Figure 6.13.

SUMMARY

- Controlling dynamic range determines the compression of the echo signal and how many grey levels may be used to display the image.
- A smaller dynamic range will make the image appear more 'contrasty' and less smooth. Fine degrees of detail will be lost.
- The greyscale map used determines how the range of echoes, within the dynamic range, are distributed across the greyscale displayed.
- By using a non-linear mapping, the majority of the greyscale may be used to display the clinically useful information. Low-level noise and high-level bright echoes are de-emphasised.
- When using an 'S'-shaped mapping, the steepness of the curve will make the image more or less contrasty, with the contrast cut-off level determined by the positioning of the steep gradient within the dynamic range.

CONTRAST RESOLUTION

Contrast Resolution is the ability to detect a target whose grey level is different from the surrounding image grey level. It is discussed in Chapter 7.

USER CONTROLS

Many scanners enable the user to control the **dynamic range** and which **mapping** is used. The level to set the dynamic range and the mapping to choose depends on a number of factors. Some scans benefit from a more 'contrasty' image, e.g. some vascular scans or scans where it is important to see boundaries and interfaces within tissue. Other images benefit from a 'smooth' image with a wide dynamic range, e.g. looking for subtle changes in a diseased liver or when looking for fine detail.

As already noted, the **main signal gain** interacts with the dynamic range and must be adjusted to optimise the visualisation of clinically useful information in the image.

Some sonographers have a natural preference for a more or less contrasty image. It is the case that your 'eye-brain' does get used to the image it regularly sees and will pick up the necessary information from an image that another sonographer would find less helpful. This partly explains why one person prefers the image of one scanner over another – you get used to what you are familiar with. However, *it is very important to use a dynamic range and mapping that does not hide or minimise the information you need to see for a particular scan.*

One of the things a **machine preset** (see below) for a particular scan type does, is to set a dynamic range and greyscale map for that scan. It is important for you as the sonographer to ensure that the dynamic range and mapping is suitable for *the patient in front of you* and to adjust them if necessary. A machine preset is not necessarily the optimal setting.

IMAGE MEMORY

The image memory acts as a **scan converter** that converts an image from one format to another. For example, write to the memory in a sector format, addressing pixels to match acquisition of scan lines and then read out the memory in a rectilinear video display format at video rate (Figure 6.14).

FRAME FREEZE

The image may be frozen to examine it, make measurements and move to long term storage.

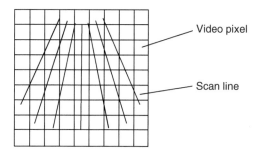

FIGURE 6.14 Map of the image memory showing the relationship between the acquired image lines and the rectilinear display format sent to the viewing monitor.

NOTE

Beware of introducing blurring in an image by moving the probe as you press 'freeze'.

READ AND WRITE ZOOM

It is useful to be able to zoom in on a part of the image. Depending on how it is done, it may or may not give more information. The zooming may be done as **read zoom** or **write zoom** (Figure 6.15).

Read Zoom

The image is frozen and stored in memory and part of that stored image is read out and enlarged to display on full screen. No new echo information is produced but fine details may be clearer, e.g. for making measurements.

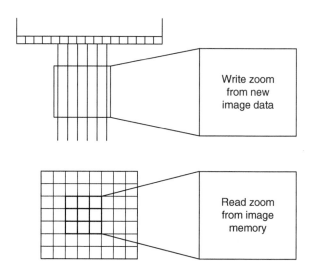

FIGURE 6.15 Illustrating the difference between write zoom which uses new scanning data and read zoom which expands the image stored in memory.

Write Zoom

Write zoom uses new echo information to form the zoomed image. Instead of scanning a full field, the ultrasound scanner just sends pulses into a restricted field of view. This restricted field of view is then used to fill the whole display. Possible benefits are:

- Increased frame rate as there are fewer lines to interrogate.

or

- More closely spaced scan lines into the smaller field of view. This preserves frame rate but reduces the need to interpolate so much (see Chapter 7). When displayed, this 'zoomed' area has more real information as more pulses were used to cover the field of view.

Remember – **read zoom** displays enlarged frozen data, **write zoom** displays live data at a larger scale.

IMAGE PROCESSING

A number of image processing techniques may be applied to the image data. These are aimed at improving the quality or appearance of the image and to increase the diagnostic information available by bringing out certain features in the image.

They may be applied as pre-storage or post-storage processing. That is, whether the processing is performed on the data before it is stored in memory or whether it is performed on the stored data.

Image processing is fully discussed in Chapter 7 and section "Advanced Image Processing" in Chapter 12.

USER CONTROL

Exam Presets

The manufacturers of scanners provide the user with presets for the examination of particular clinical areas of the body, e.g. small parts, abdominal, foetal exam etc.

> **NOTE**
>
> When activated each preset will change a wide range of controls including power output, grayscale mapping, frequency, image processing etc.

In addition to presets provided by the manufacturer there is usually the option to set up your own presets and store them for future use. **It is important that as a user, you know how to adjust the scanner controls to give a clinically useful image *yourself,* without relying on the presets.** Presets, whether the manufacturers or your own, provide a quick way to progress an examination but you should take the time to find out what each preset is adjusting and then make your own adjustments in the light of that information. It will also be necessary to make adjustments during an examination so that your machine is giving the optimal image for the patient in front of you.

Measurements

Using the image stored in the memory, cursors may be superimposed to make measurements. Measurements are discussed in detail in Chapter 10.

Figure 6.16 shows a block diagram of a full B-mode scanner.

NOTE

The order of some processes may differ on different machines.

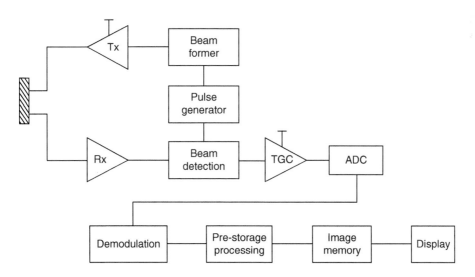

FIGURE 6.16 Block diagram of a complete B-mode imaging system.

CHAPTER 7

Image Quality and Artefacts

As with all medical imaging, image quality is an important issue in ultrasound imaging. This can be divided into the following:

- Image resolution (see Chapters 3 and 5).
- Perceived appearance – image smoothness, contrast, etc.
- The presence of artefacts within the image.

Resolution is the ability to image fine detail spatially or temporally (fast events). Apart from the intrinsic resolution that a particular beam and pulse shape can give, related to frequency and focusing, a number of other factors affect the resolution seen. The **intrinsic resolution** is the best resolution achievable from a given transducer with a chosen set up of frequency, focus, and processing. Other factors such as tissue type and patient movement will tend to degrade what is actually resolved in the image.

ACOUSTIC WINDOW

The position of the probe on the skin and the angle of approach to the target of interest defines what is called the **acoustic window**. The choice of acoustic window can make a very big difference to the quality of the image seen, e.g. avoiding excessive overlying fatty tissue or gas filled gut when observing the abdomen. Some approaches

Ultrasound Technology for Clinical Practitioners, First Edition. Crispian Oates.
© 2023 John Wiley & Sons Ltd. Published 2023 by John Wiley & Sons Ltd.

to a target provide a particularly good acoustic window, e.g. using a filled bladder as an acoustic window when observing the pelvic region.

FRAME RATE: FRAMES PER SECOND (fps)

Frame rate is the rate at which the ultrasound image information is refreshed. The display shown on the screen is refreshed many times a second so as to display a continuous image. However, depending on a number of factors, the ultrasound information being displayed may not be acquired fast enough so as to show no flicker or 'joltiness' when imaging very fast moving structures such as the heart valve leaflets.

The rate at which new images can be acquired determines the **temporal resolution** – the separation of events in time.

The human eye has a temporal resolution of ~40 ms which equates to 25 fps (Recall: frequency = 1/time)

An image with <20 fps will show '**image flicker**'.

What Determines Frame Rate?

- The number of ultrasound lines/beams in a frame
- The depth of image. The next pulse cannot be sent until echoes from the last pulse return from the greatest range.

$$\text{Time for one image line } \Delta t = \frac{2d}{c} \quad c = 1540\,\text{ms}^{-1}$$

So, $\Delta t = 13\,\mu\text{s}\,\text{cm}^{-1}$ ('go and return')

100 lines to 10 cm depth $= 100 \times 130 = 13000\,\mu\text{s}$ so frame rate $= 76$ fps max

We need to leave a gap for echoes die away before the next pulse is transmitted, so frame rate will always be lower than the maximum possible.

As seen in Figure 7.1, a sector probe will need more lines to fill in gaps distally as beam directions spread out (although distal line density need be no more than linear array line density).

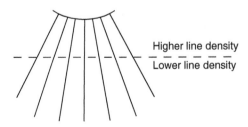

Higher line density

Lower line density

FIGURE 7.1   Reduced line density in the far field of a sector scan format.

Example: 5.5 cm linear array with focused beamwidth of 0.4 mm needs 137 beams to cover the width of the array field of view with non-overlapping beams.

A typical linear array has 160 elements.

This could yield a maximum frame rate of ~100 fps but in practice frame rate will be significantly lower as a result of multi-zone transmit focusing, compound imaging, etc.

Very much higher frame rates can be achieved using **synthetic aperture techniques** (see Chapter 13).

There are a number of ways that frame rate can be improved, which also improve the appearance of the displayed image.

Where temporal resolution is critical, e.g. in cardiac work, then it is important to minimise the number of pulses required to obtain an image. The ways in which this can be done are as follows:

- Reduce the size of the displayed field, e.g. reduce the sector width in a sector scan. This reduces the number of scan lines required to cover the field.
- Reduce the depth displayed to what is necessary. This reduces the 'go and return' time for each pulse.
- Use a single transmit focus set to the depth of the target of interest. Multiple transmit zones require a separate transmit pulse for each zone.

In addition to adjusting these parameters, the machine may use **interlacing** of scan lines and **interpolation** to reduce the actual number of transmitted scan lines.

INTERLACING SCAN LINES

The line density in an image can be increased by the use of interlacing (Figure 7.2). To form an interlaced image **frame**, the field of view is covered in two sweeps or **fields**. In the first field, beams are sent out along the odd numbered lines. In the second

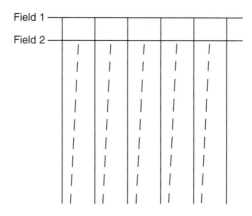

FIGURE 7.2   Illustration of interlacing with the field 2 angled with respect to field 1.

field, beams are sent out along the even numbered lines. The lines in the second field may be angled very slightly with respect to those of the first field to 'fill in the gaps'. A complete displayed **frame** is then made up of two interlaced fields that are both stored and displayed as one image. It has the advantage that each field only requires half the number of beams, although there will be some blurring (loss of temporal/spatial resolution) for fast moving structures where movement has occurred between the two fields being acquired. Because the two fields are angled slightly to one another, the speckle pattern from each will be different and the image will appear smoother. A further development of this idea is compound imaging (see below).

INTERPOLATION – WRITING IN 'EXTRA LINES'

Frame rate can be improved at the expense of resolution by using interpolation (Figure 7.3). This involves transmitting a smaller number of scan lines and filling in the gaps by creating synthetic lines (i.e. no new data). Interpolation has been used to fill out sector images as the line density reduces distal to the transducer.

The pixels in the inserted lines may be an average of the adjacent real line data or may use a more sophisticated algorithm taking into account local area brightness, emphasising edges, etc.

Pros
- Improves frame rate
- Improves image appearance cosmetically

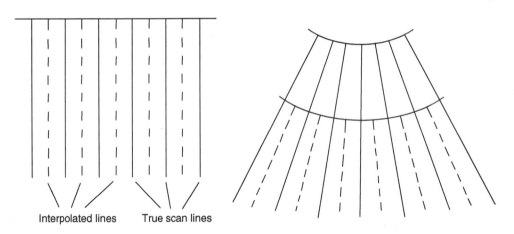

Interpolated lines   True scan lines

FIGURE 7.3   Interpolated lines generated between echo generated lines for a linear format and a sector scan.

Cons

- The interpolated lines do not contribute new data about the tissue being examined.
- There is a reduction in resolution within the image as some information from between the scan lines is missing, even though the ultrasound beams may be highly focused to be narrow throughout their range.

SPECKLE

Speckle results from the fact that ultrasound imaging uses **coherent waves** – it is a **coherent imaging system**. This means that the phases of the sound waves that form the ultrasound pulses are all aligned such that peaks and troughs of the waves can interact through superposition to create maxima and minima in the ultrasound beam and produce speckle in the echoes (see Chapter 2).

Because the speckle pattern depends on the microscopic positioning of reflectors smaller than the wavelength of the ultrasound, it is a random noise on the image where the peaks and troughs appear in different positions each time the tissue moves or is interrogated by an ultrasound beam from a different direction.

Question: Is Speckle an Artefact?

- It is the way tissue is seen on the image.
- It is a random speckle pattern, not an image of microscopic structure.
- Different tissues will have different speckle patterns depending on their microscopic structure, e.g. normal liver versus metastases, so the speckle pattern seen does give information on the underlying nature of the tissue.
- Speckle can also be considered to be noise against which we want to see truly resolved targets.

A number of techniques are used to reduce speckle. These are as follows:

- Frame averaging
- Compound imaging
- Adaptive filtering techniques applying more complex algorithms to smooth speckle

The ideal would be the following:

- To see edges of organs and structures large enough to be resolved including changes in tissue parenchyma, e.g. metastases
- To smooth the image where targets are too small to be resolved

FRAME AVERAGING OR PERSISTENCE REDUCES SPECKLE IN THE IMAGE ('GRAININESS')

In order to smooth over the random speckle seen within each frame, a percentage of the previous frame values are added to the current frame on a pixel by pixel basis (Figure 7.4). For example, 20% of each pixel value from the last frame is added to the present frame. The speckle will change with the minute movements of the transducer from frame to frame and so average out, whereas real structures in the image will not change (Figure 7.5). The use of a proportion of previous frames is clearly seen in Figure 7.27, where the image of each of the frames is spread out in time.

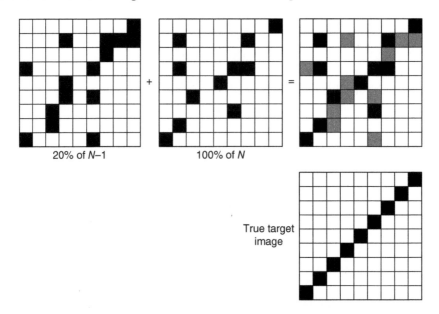

FIGURE 7.4   Diagram showing the principle of frame averaging to reduce the effect of speckle.

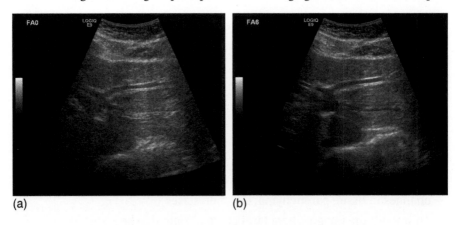

FIGURE 7.5   Showing the effect of frame averaging. No frame averaging in (a) and averaging six frames (b). Note how the speckle in the image is smoothed.

USER CONTROL

The degree of **frame averaging** may be selected by the user. The more frame averaging used the greater will be the loss of temporal resolution for moving targets. If a lot of frame averaging is used (high percentage over several frames), then there will be some noticeable blurring of the image as the probe is moved on the patient's skin. It is important to ensure the target image is as stationary as possible before freezing the image.

In addition to the speckle pattern not directly displaying microscopic information about the tissue, there is frequently a more general degredation of image quality caused by a small amount of image data appearing in the wrong place within an image. This goes under the generic name of **clutter**.

SPATIAL COMPOUND IMAGING

Curved surfaces are poorly visualised as the surface curves towards the direction of propagation.

Using beam steering, compound imaging sends multiple beams (three to nine beams) at different angles into the scan plane so more edges are perpendicular to the beam (Figure 7.6). This gives better visualisation of curved surfaces. It also acts to smooth speckle and clutter in the image.

NOTES

- The frame rate is not greatly affected. Acquisition of the images from each direction is in the form 1–2–3, 1–2–3, for example. The scanner stores the image from each transmitted direction and combines it with the images from the other directions dropping the oldest and retaining the newest with each field sweep.
- The image displayed is the sum of the echoes from beams from multiple directions.
- Compound imaging also smooths out speckle as it averages the speckle from multiple beams cf. frame averaging.
- For the same reason, it also reduces clutter.
- There may be some loss of resolution if the speed of sound along the different directions of the beams is not uniform, leading to some blurring of the image (see section 'ARTEFACTS').
- Activating compound imaging is a user control.

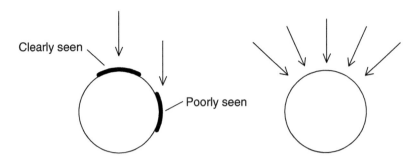

FIGURE 7.6   The improvement in visualisation of curved surfaces gained by using compound imaging.

ADAPTIVE FILTERING

Adaptive filtering and other advanced image processing algorithms aim to enhance real structures in the tissue and reduce speckle and noise. These algorithms may include edge enhancement.

Edge Enhancement

Edge enhancement aims to make real edges in the field of view visible and continuous on the image whilst suppressing spurious echoes that may look like edges.

As an example, one such adaptive filter performs an analysis of the image in one step and then then applies the filtering in a second step (Figure 7.7). The filtering is called 'adaptive' because it adapts what it does depending on what it finds in the analysis stage. It endeavours to identify real structures both within an image and across images from frame to frame. As well as trying to detect real structures and artefacts, it uses texture properties and statistical analysis of the images at several scales. In the enhancement phase, it applies edge enhancement for interfaces and smoothing of uniform areas to suppress speckle and noise [1].

Such filters may be activated as a user control.

CONTRAST RESOLUTION

This is an important concept that describes the ability to detect one target against another, e.g. the ability to detect a subtle change due to a tumour in normal tissue.

Contrast resolution is partly subjective and depends on the following:

- The size of the target
- The contrast difference between the target and the surrounding image
- How 'noisy' the image is, including speckle, clutter, and poor visualisation
- Greyscale mapping and dynamic range used

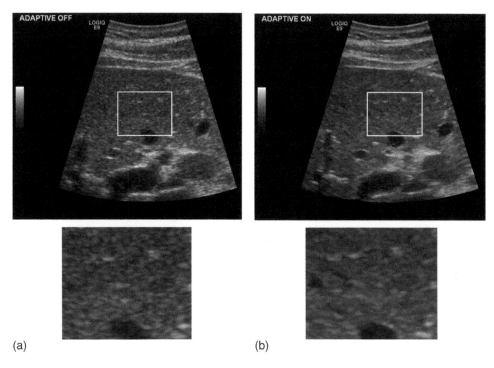

FIGURE 7.7   Showing the effect of applying one example of an adaptive filter. Filter off (a) and filter on (b). Note the change in the appearance of the speckle pattern as shown below each image.

- Human visual perception
- Viewing screen set-up
- Background lighting in the room

Noise in an Image

Noise in an image will reduce the ability to visualise and detect targets of interest. Sources of noise include the following:

- Speckle
- Poor visualisation due to weak echoes at depth. At the limits of penetration, the electrical noise in the system has a similar amplitude to the echo signal, as seen when the gain is set high.
- Poor visualisation due to beam distortion from overlying tissue, e.g. some types of fatty tissue.
- There will be clutter in the image if echo information from side lobes, grating lobes and multiple reflections is present.
- Probe movement at the moment of freezing an image will degrade the image and hence detectability of targets.

The **greyscale mapping, dynamic range,** and **gain** chosen will also affect contrast resolution. In particular, the choice of mapping and dynamic range can make the image more contrasty or smoother in its display of grey levels. This was discussed in Chapter 6.

Tissue mimicking phantoms that have calibrated changes in greyscale targets may be used to demonstrate that no subtle changes are being missed in a particular mapping setting.

Colour Enhancement of Greyscale Images

In addition to choosing a greyscale mapping to use, the user may be able to 'colour up' the image. A single colour 'greyscale' is used instead of a black and white greyscale. In particular, some scanners allow a colour enhancement to be used that matches the visual sensitivity of the human eye within the optical spectrum (Figure 7.8). The eye is most sensitive around a yellow–green colour.

Human Visual Perception

- The human eye is better at detecting changes of brightness when the change is across a sharp boundary than when the change is gradual.
- Smaller targets require a bigger change of contrast to be seen than larger targets.
- The eye detects horizontals and verticals better than oblique changes of brightness.
- With sharp bright illumination, we can detect grey levels that differ by 1%.

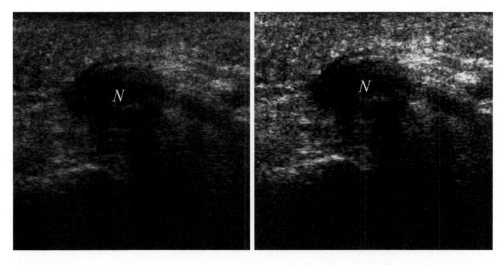

FIGURE 7.8   Example of a Morton neuroma showing the effect of colouring the B-mode image to match the maximum sensitivity of the eye. *Source*: Sofka et al. [2]/with permission of John Wiley & Sons.

> **NOTE**
>
> In relation to perceiving target size against speckle, viewing distance and magnification (zoom) will affect detectability.

How to Set Up a Monitor

Before attempting to view and interpret clinical images, the monitor on which they are viewed should be properly set up for optimal viewing. For a greyscale image, the monitor may be set up using the greyscale wedge display usually displayed at the side of B-mode images. The monitor brightness and contrast should be adjusted so that peak white shows as peak white and black has no illuminance at all. Every step in the wedge should be visible as a step. In particular, levels near the black and white ends should be distinguishable. A starting point is to set the contrast high and the brightness low and then increase the brightness until all the steps are seen.

In order to see some of the factors that affect viewing on a monitor, there are many web sites that can be viewed.

Image Flicker

Image flicker on the monitor itself is not really a problem most of the time, as the displayed image can be re-read to the display from image memory at least 25+ times/s. However, a low frame rate will lead to a jumpy image as the transducer is moved on the patient's skin. It may be also be seen as image blurring as the probe is moved, e.g. too much frame averaging (persistence).

Ambient Lighting

It is important to recognise the effect ambient lighting in the scan room has on the ability to see fine detail in displayed images. The position of the lighting may also be important, e.g. if a light source reflects directly from the viewing screen surface or off glasses you are wearing, causing flare, the contrast resolution seen in the image is reduced.

There is an eye–brain adaption to the average brightness of a scene being viewed. For example, the eye–brain adjusts sensitivity when going from bright sun into a darkened room. This **dark adaption** takes time as there is a physical adaption (the iris) and a biochemical-neural adaption. To become fully dark adapted on a dark night when looking at stars takes about 20–30 minutes.

The contrast between the average screen brightness and the level of ambient lighting should therefore be considered when scanning.

ARTEFACTS

DEFINITION

An artefact is any feature in an image that is not an exact or accurate one-to-one representation of the target being imaged.

Artefacts need to be clearly recognised and understood by the sonographer so as to avoid mis-interpretation of the images. Some artefacts can be useful.

Assumptions

Ultrasound imaging is based on the following assumptions:

1. Speed of sound is constant in the body.
2. Attenuation in tissue is constant and uniform throughout the image.
3. The beam axis is straight throughout the range of the beam.
4. Later echoes result from targets at greater distance from the probe.
5. The ultrasound beam is infinitely thin, i.e. no echoes arrive that are away from the beam axis.
6. The image is acquired instantaneously.

Artefacts will occur when these assumptions are not met. They can be considered around these basic assumptions.

SPEED OF SOUND ARTEFACTS

The assumption is that the speed of sound, $c = 1540\,\text{ms}^{-1}$, but it does vary between soft tissues by $\pm 5\%$

Refraction

Refraction leads to distortion of the image (see Chapter 2)

Refraction results from changes in the speed of sound at interfaces where the beam is not perpendicular to the interface. The machine does not know the beam has changed direction and distal targets will be misplaced on the image, distorting the structures seen (Figure 7.9). This artefact involves assumptions 1 and 3 mentioned above.

Speed of sound differences with in the tissues will cause some degradation of the image when **compound scanning** is being used. The effect of this is apparent in

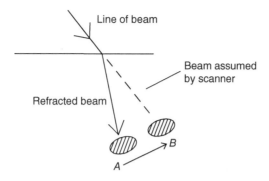

FIGURE 7.9   Diagram showing how refraction causes distortion in the image by misplacing targets, e.g. from *A* to *B*.

a scan of a phantom with an inclusion having a different speed of sound as seen in Figure 7.10. Note the appearance of the wires in the image. However, such degradation will be generally the case across the whole clinical image. Figure 13.12 shows the effect of correcting for speed of sound on this phantom.

Examples:

- Poor visualisation of deep tissue due to distortion of the beam by overlying fatty tissue (Figure 7.11) (see Chapter 2).
- A biopsy needle may appear bent as it crosses a tissue boundary where the speed of sound changes, like a pencil appearing to be bent when put in a glass of water, where the light moves more slowly in water than in air causing refraction of the view of the pencil (Figure 7.12).
- Lensing may be seen in a cardiac scan where a dense structure causes the ultrasound to be refracted to give a double image of a reflecting target (Figure 7.13).

Axial Misplacement

Axial misplacement (caused by assumptions 1 and 4 not being the case)

The example shown in Figure 7.14 shows the error in measuring the diameter of a blood vessel due to axial misplacement.

Blood vessel	In blood	$c = 1570\,\text{ms}^{-1}$
	Machine assumes	$c = 1540\,\text{ms}^{-1}$
	Difference	1.9%

So, *A–P* diameter has ~2% error in measurement.

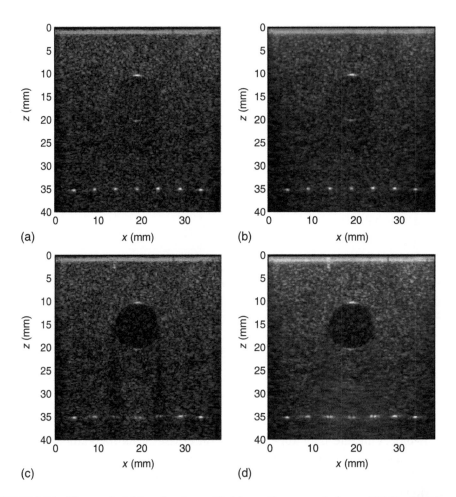

FIGURE 7.10   Tissue mimicking phantom with (a) a uniform speed of sound (SoS) of $1520\,\text{ms}^{-1}$ imaged from a single forward angle and (b) with compounding. A phantom with an inclusion having a different SoS ($1470\,\text{ms}^{-1}$) imaged from a single forward angle (c) and with compounding (d). *Source*: Jaeger et al. [3]/IOP Publishing/CC BY 3.0.

Axial misplacement may also be seen when the beam passes through a tissue, such as a lipoma, where the speed of sound is slower than $1540\,\text{ms}^{-1}$. The posterior structures will then be placed more distally on the image than they should be (Figure 7.15).

As we have seen, differences in speed of sound can become a problem when multiple beams are used to image a target from several directions, such as in compound imaging. If we could use the true speed of sound for each tissue layer, image quality could be considerably improved. A first-order correction to the speed of sound problem is to choose a different speed of sound for the scanner to assume for the **system velocity** and see if that improves resolution in the image. In other words, to use a different average speed of sound than $1540\,\text{ms}^{-1}$. This can improve image quality as the ultrasound paths from different elements towards a focus will be better matched to the average speed of sound thereby improving the focus and hence image

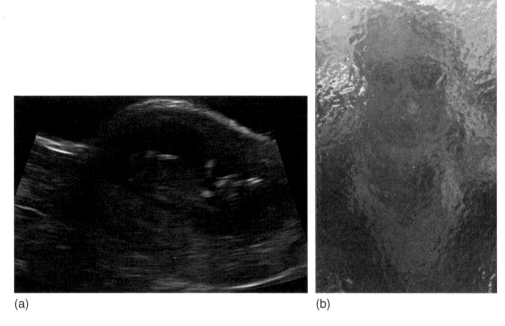

(a)   (b)

FIGURE 7.11   Poor visualisation due to distortion of the ultrasound beam by overlying tissue (a) is similar to the distortion of an image caused by 'wobbly' glass (b). *Source*: Oates and Taylor [4]/with permission of SAGE Publications.

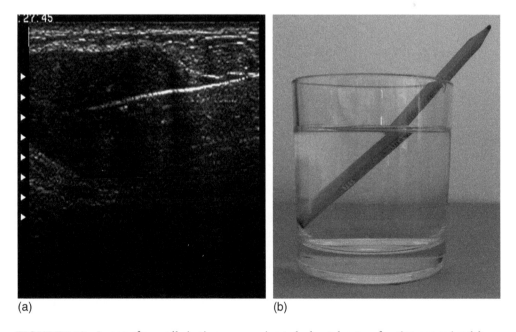

(a)   (b)

FIGURE 7.12   Image of a needle in tissue appearing to be bent due to refraction occurring (a). *Source*: Courtesy Tim Lees. (b) Optical refraction for comparison.

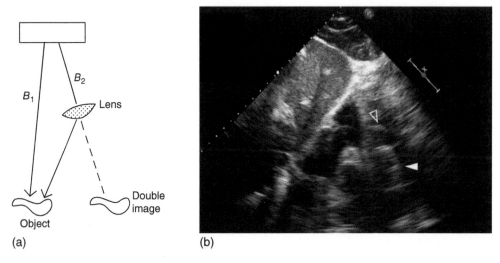

(a)   (b)

FIGURE 7.13 Lens artefact. B_1 is the direct beam. For beam B_2, the ultrasound is refracted towards a target to produce a second image of the target along the line of refraction. The example shows a double image of the aorta (full arrowhead) in a subcostal short-axis image of the heart, due to refraction of the ultrasound beam at perihepatic fatty tissue (red arrow). A Swan–Ganz catheter in the right ventricular outflow tract is doubled as well (empty arrowhead). *Source*: Bertrand et al. [5]/with permission of Elsevier.

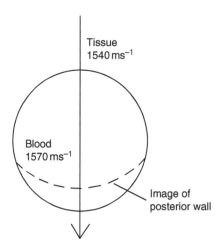

FIGURE 7.14 Diagram showing the misplacement of the distal wall of a blood vessel due to speed of sound error when the machine assumes $c = 1540\,ms^{-1}$.

quality. One technique to achieve this is to try a different speed of sound and examine the image to see if edges are more sharply defined. A number of speeds may be tried to find the best solution and different speeds may be used in different regions of the image. This technique to improve speed of sound correction has been used and automated in some scanners [6].

Full speed of sound correction is discussed in Chapter 13.

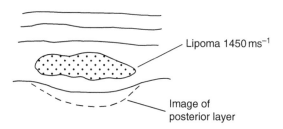

FIGURE 7.15 Diagram showing axial misplacement of distal tissue due to speed of sound in lipoma being less than $1540\,ms^{-1}$.

ATTENUATION ARTEFACTS

The scanner assumes a constant attenuation coefficient such that when TGC is correctly adjusted, the image has uniform brightness throughout the field of view.

Poorly Adjusted Time Gain Control (TGC)

Poorly adjusted time gain control (TGC) can cause information to be missed due to insufficient gain or due to excessive gain (see Figure 6.4).

Acoustic Shadowing

Acoustic shadowing is caused by structures that are strongly reflective or have higher attenuation than surrounding soft tissue. It is the acoustic analogue of optical shadowing such as would be caused by a solid object blocking the beam from a torch. Distal tissues are not insonated and appear dark on the image.

Acoustic shadowing may be caused by the following:

- Hyper-echogenic targets, e.g. bone, gas, calcification reflecting or absorbing nearly all the sound impinging upon them (Figure 7.16).
- Oblique surfaces of targets, e.g. arterial wall in cross section, where the beam is deflected away from distal targets so they are not insonated (Figure 7.17).

NOTES

- Shadowing restricts viewing angles, e.g. ribs and bowel gas, when distal tissues need to be seen.
- Shadowing may be useful as it shows a solid acoustically opaque target.

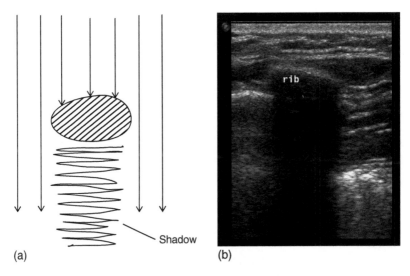

FIGURE 7.16   Shadowing caused by an acoustically opaque target, in this case a rib.

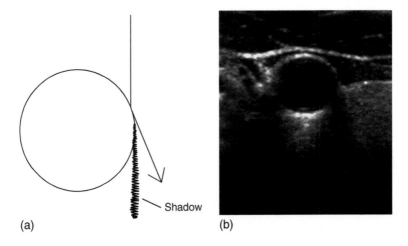

FIGURE 7.17   Transverse view of the right neck showing shadowing due to the ultrasound beam hitting the common carotid artery wall at an oblique angle and being deflected into adjacent tissue.

Post Cystic Enhancement

Post cystic enhancement is caused by the attenuation being lower than expected (Figure 7.18).

Beam A in solid tissue has TGC applied to give a uniform image.

Beam B passes through non-attenuating fluid. TGC increases the gain of the echoes throughout the range of the beam, but as there is almost no attenuation in the fluid, distal targets have a stronger pulse hitting them. Their stronger echoes are then amplified too much and show up brighter than adjacent targets from beam A, hence **post cystic enhancement**

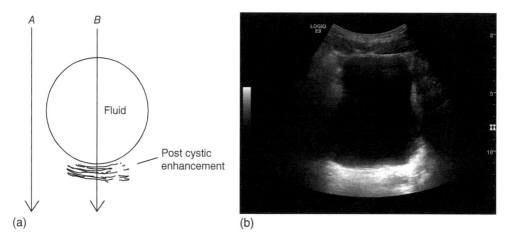

FIGURE 7.18   Post cystic enhancement caused by the passage of ultrasound through a low attenuating fluid, in this case the bladder.

> **NOTE**
> This is a useful artefact – it tells you the structure is fluid filled and is not solid.

Shadowing from the Transducer Surface

Shadowing from the transducer surface results from poor contact with body, e.g. lack of gel on probe/skin. This can be identified as a shadow that extends throughout the image from the probe-skin interface (Figure 7.19).

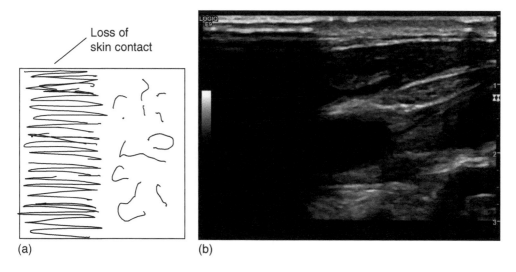

FIGURE 7.19   Absent echoes from the transducer surface due to poor acoustic contact at the skin.

REFLECTION ARTEFACTS

Reflection artefacts result from assumptions 3 and 4 above, not always being the case.

Mirror Image Artefact

Mirror image artefact may occur where there is a plane reflecting surface (Figure 7.20)

The scanner assumes the beam is straight – echoes from R are therefore shown at T on the image.

Examples of reflection artefact may be seen in the lung when the probe is angled cephalad towards the diaphragm (Figure 7.21). Liver echoes may be seen in the lung, a gas filled space that should have no echoes within. In a second example, the subclavian artery may be seen reflected in the upper surface of the lung (see Figure 9.20).

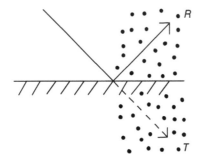

FIGURE 7.20 Diagram showing how reflection at an interface can cause targets R to be misplaced on the image at T to give reflection artefact.

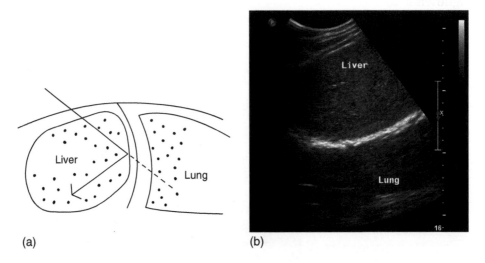

(a)　　　　　　　　　　　(b)

FIGURE 7.21 Cephalad angulation onto the diaphragm through liver. Liver echoes are seen in the lung space.

Reverberation

Multiple reflections between parallel surfaces (these artefacts involve assumption 4).

Reverberations can show themselves in several forms creating a number of named artefacts.

Reverberation Artefact

Reverberation artefact may often be seen when two or more interfaces lie parallel to one another (Figure 7.22). When the beam is perpendicular to these surfaces, the pulse can reverberate between these surfaces producing a string of separate echoes back to the probe. As these arrive at a later time than the first echo, they are positioned on the image at greater depth. For two parallel surfaces, the second surface is reproduced at the same depth below the first surface as the distance between the two surfaces, since the pulse has travelled the same distance a second time. The reverberation echoes get progressively fainter as the sound is attenuated along its multiple echo path. An example is shown in Figure 7.23.

Probe – Interface Reverberation

Reverberation between a bright reflecting target and the probe surface gives a second echo at twice the depth of the target $(2 \times d)$, where d is the target range. This is easily detected – move the probe up and down on the skin and the reverberation echo will move twice as fast as adjacent true echoes (Figure 7.24).

Ringdown

Ringdown is caused by reverberation within the front face of the probe, i.e. within the lens and matching layer, and appears as a thin region of bright echoes immediately adjacent to the probe-skin surface on the image (Figure 7.25). With the use of good matching layers and low impedance materials in the probe face, this artefact is greatly reduced.

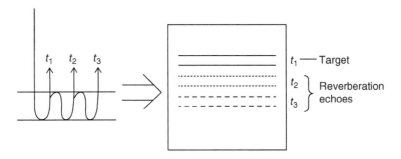

FIGURE 7.22 Diagram showing how reverberation artefact is generated by multiple reflections from parallel surfaces.

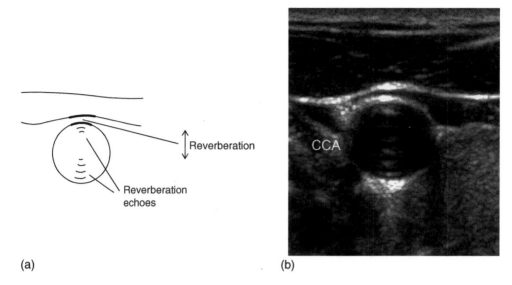

(a)   (b)

FIGURE 7.23 Reverberation seen within the lumen of the carotid artery arising from multiple reflections at the proximal wall.

Comet Tail or Ringing

There are two separate causes that give a similar appearance on the image.

- Multiple reflections within a solid target. The speed of sound within the target is high and the size of the target is such that the multiple echoes do not show up as individual echoes but merge to produce a streak or train of echoes in the image (Figure 7.26).
- Small air bubbles or fluid trapped between air bubbles may be caused to resonate when hit by the ultrasound pulse. They then go on ringing as a bell would when hit, producing a string of echoes that appear as a streak in the image.

NOTES

- Comet tail is useful for detecting foreign bodies in tissue, e.g. IUCD and wires/staples.
- Small bubbles of air show as bright echoes with streaks in the image. This may be useful, e.g. to see where a needle introducing fluid is placed. It can also be a problem when it blocks viewing distal tissues.

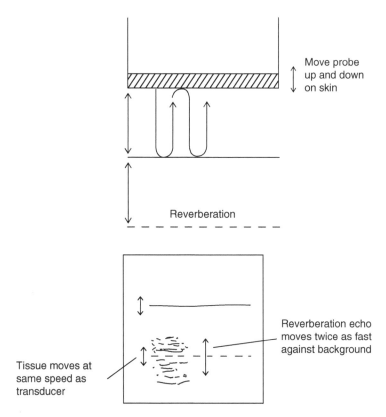

FIGURE 7.24 Reverberation artefact arising from reflection at the transducer surface and how to detect its presence by moving the transducer up and down on the skin surface.

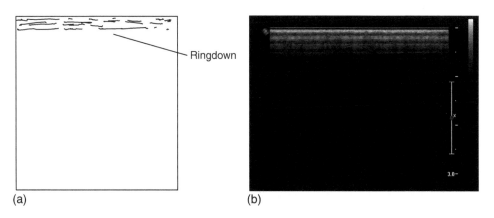

FIGURE 7.25 Ringdown from the front face of the transducer seen with the transducer in air. When in contact with the skin this ringdown will be less apparent. *Source*: Courtesy of Barry Ward.

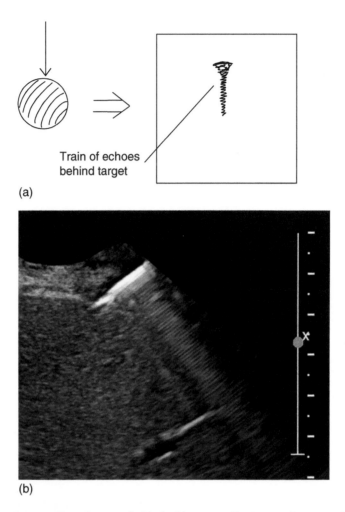

Train of echoes
behind target

(a)

(b)

FIGURE 7.26   Comet tail artefact seen behind a biopsy needle. *Source*: Courtesy of Tim Lees.

ANISOTROPY

Anisotropy describes the case when a tissue changes its appearance, scattering sound differently, as the orientation of the target with respect to the scan plane changes. The tissue is then said to be **anisotropic**.

Some tissues such as muscles and tendons have a filamental structure with strongly oriented cells or fibres. These can act in concert together to alter the echogenicity of the tissue depending on the angle of insonation (Figure 7.27). Changing the orientation of the beam from along to across the axis of the tissue fibres may change the appearance of the tissue from a hyperechogenic tissue to a hypoechogenic tissue. A change of just 5° in angulation of the beam may be sufficient to produce this change. This is a reflection artefact as the reflectivity of the tissue changes with orientation of the ultrasound beam.

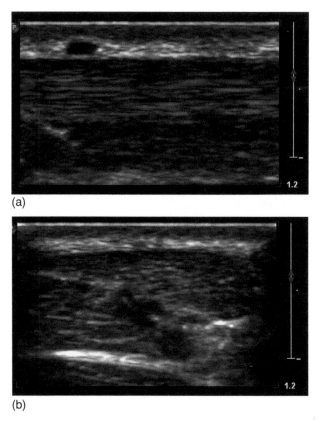

(a)

(b)

FIGURE 7.27   Anisotropy seen in a superficial muscle imaged (a) longitudinally and (b) transversely. Note the different speckle pattern texture seen in each case.

> **NOTES**
> - Mis-diagnosis is possible, e.g. a normal tendon may appear damaged or ruptured.
> - Tendons may be differentiated from nerves, which do not show anisotropy.
> - Examine potential anisotropic targets from more than one direction.
> - Compound imaging will reduce the effect of anisotropy as the image is formed using beams from several directions.

BEAM SHAPE ARTEFACTS

Beam Shape Artefacts – caused by assumption 5 not being the case

Slice Thickness Artefact

Slice thickness artefact is caused by inclusion of targets outside the image plane but within the elevation beamwidth. This results from the fact that the beam and scan plane is not infinitely thin. An example from within an artery is shown in Figure 7.28.

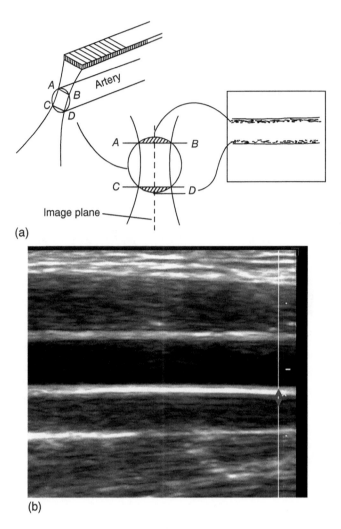

(a)

(b)

FIGURE 7.28   Slice thickness artefact seen within the brachial artery imaged in longitudinal section. The vessel wall outside the scan plane is still within the slice thickness of the beam.

Side Lobes and Grating Lobes

Side lobes and grating lobes: These generally degrade the detail in the image as echoes come from other directions than the beam axis and are added to the beam image. They may be seen in cystic structures such as the bladder where diffuse echoes may be seen within the cystic fluid. They are also specifically seen in echocardiographic scans where a strong reflector is seen stretched out into a space beyond its real position (Figure 7.29).

Slice thickness artefact, grating lobe, and side lobe artefact will be present to some degree in all images. They contribute to 'clutter' in the image and generally degrade the image seen. However, it is important to recognise them as artefacts in specific situations where they are seen, such as those indicated in the examples given above.

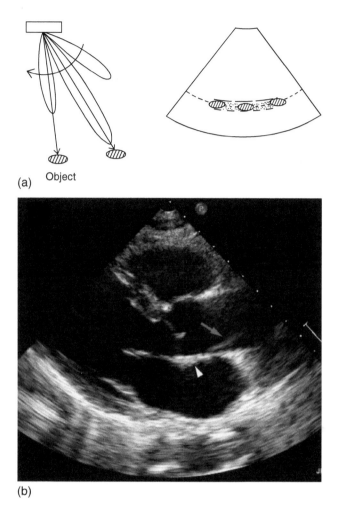

(a)   Object

(b)

FIGURE 7.29   Echoes from the side lobe register along the main beam direction, then the main beam and opposite side lobe in turn pick up echoes to give a streaked appearance for the object. Example shows cardiac parasternal long axis view with linear side lobe artifact (arrow) in the ascending aorta due to a calcified sinotubular junction (arrowhead). *Source*: Bertrand et al. [5]/ with permission of Elsevier.

TEMPORAL ARTEFACTS

Result from the fact that the image is not obtained instantaneously (assumption 6).

If the frame rate is low then fast moving targets will not be adequately tracked, they will move in the time it takes to acquire one frame. Artefacts will result, as may be shown using a rotating wire phantom (Figure 7.30a). A single straight line of wires crosses the image plane and when stationary appears as in Figure 7.30b. When the phantom is rotated, the image appears as in diagram Figure 7.30c. In the image shown

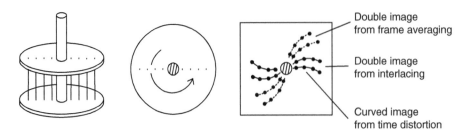

FIGURE 7.30   Diagram of a moving phantom to show temporal artefact.

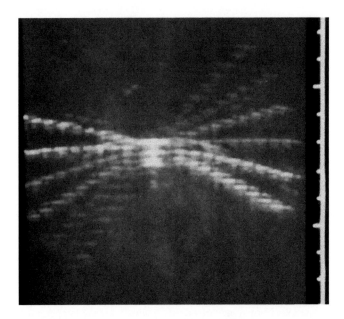

FIGURE 7.31   An image taken from the moving phantom shown in Figure 7.30. Duplication giving image pairs is due to interlacing of fields. Successive pairs are due to frame averaging. Each line of echoes is curved due to movement over the time taken to acquire the image.

in Figure 7.31, there are three sets of double images. The pairing is caused by interlacing of two image fields and each line of wires is not straight but is distorted due to movement of the target during the time the frame was acquired. The three sets of images are the result of frame averaging. Each set is weaker than the previous set as the proportion of its contribution to the image reduces.

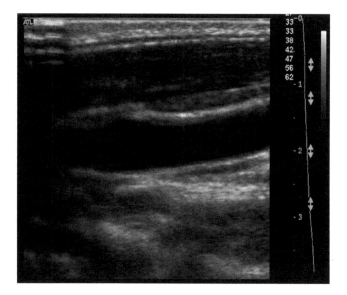

FIGURE 7.32   Image of a carotid artery showing the effect of movement of the probe during moment of freezing the image causing blurring. This was exacerbated by the use of multiple transmit focal zones and frame averaging.

Image Blurring

Use of high levels of frame averaging can cause image blurring on the live image during movement of the probe on the skin.

Image blurring on frozen images may be caused by movement of the probe at the moment of freezing the image (Figure 7.32). This may be due to hand movement by the operator or movement in the subject, e.g. patient breathing or foetal movement. This will cause loss of detail in the image and possible errors in making measurements from the image.

FINAL EXAMPLE

The final example of the need to interpret images in the presence of artefacts is a reminder that it is also important to **know the anatomy being viewed**. Figure 7.33 shows a transverse view of the lower neck. Within the larynx there are a lot of echoes. We know that the larynx is an air-filled structure and that ultrasound will not penetrate air. Therefore, all of the echoes in the normal larynx must be artefactual. This image also shows a number of other examples of the artefacts discussed.

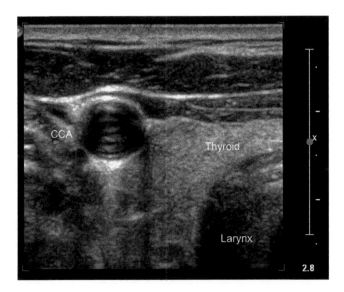

FIGURE 7.33   Transverse view of the lower neck. All of the echoes seen within this normal larynx must be artefactual as the larynx is air-filled and ultrasound does not penetrate an air interface.

REFERENCES

1. Meuwly, J., Thiran, J., and Gudinchet, F. (2003). Application of adaptive image processing technique to real-time spatial compound ultrasound imaging improves image quality. *Investigative Radiology* 38: 257–262.

2. Sofka, C.M., Lin, D., and Adler, R.S. (2005). Advantages of color B-mode imaging with contrast optimization in sonography of low-contrast musculoskeletal lesions and structures in the foot and ankle. *Journal of Ultrasound in Medicine* 24: 215–218.

3. Jaeger, M., Robinson, E., Akarçay, H.G. et al. (2015). Full correction for spatially distributed speed-of-sound in echo ultrasound based on measuring aberration delays via transmit beam steering. *Physics in Medicine and Biology* 60: 4497–4515.

4. Oates, C.P. and Taylor, P. (2016). Helping expectant mothers understand inadequate ultrasound images. *Ultrasound* 24: 142–146.

5. Bertrand, P.B., Levine, R.A., Isselbacher, M.E. et al. (2016). Fact or artifact in two-dimensional echocardiography: avoiding misdiagnosis and missed diagnosis. *Journal of the American Society of Echocardiography* 29: 381–391.

6. Napolitano, D., Chou, C., McLaughlin, G. et al. (2007). Sound speed correction in ultrasound imaging. *Ultrasonics: Supplement* 44 (Suppl 1): e43–e46. https://doi.org/10.1016/j.ultras.2006.06.061.

CHAPTER 8

Principles of Doppler Ultrasound

THE DOPPLER EFFECT

The Doppler effect is the change in the detected frequency of a sine wave when there is relative motion between the transmitter and the receiver. This will occur for all wave phenomenon. It is the change in pitch you hear when a motor bike races past you. The pitch is higher as it comes towards you and drops to a lower pitch as it moves away.

> **NOTE**
>
> For historical reasons and as it is the name of a person, Christian Doppler, who first described the effect, 'Doppler' is always spelt with a capital 'D'.

To see why there is this change in received frequency, consider the example of a boat rowing towards a source of waves on a lake (Figure 8.1). It will intercept those waves more rapidly than if the boat was stationary. If it were being rowed away from the source, the boat would intercept the waves at a lower frequency. If the boat was stationary, it would bob up and down at the same rate as the waves were being generated.

In this case, the source of waves is stationary and the boat is a moving receiver.

When a fire engine passes you at speed, with its sirens sounding, you hear a change in the pitch or frequency of the sound. The frequency is higher as it comes

Ultrasound Technology for Clinical Practitioners, First Edition. Crispian Oates.
© 2023 John Wiley & Sons Ltd. Published 2023 by John Wiley & Sons Ltd.

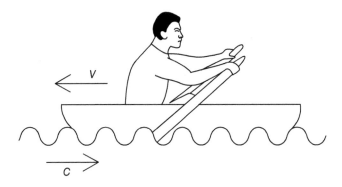

FIGURE 8.1   Illustration of a boat moving at velocity '*v*' encountering wind-blown waves moving at velocity '*c*'.

towards you and lower as it moves away. In this case, the fire engine is a moving transmitter and you are a stationary receiver.

It doesn't matter which is moving and which is stationary; it is the **relative motion** between the transmitter and receiver that causes the Doppler effect.

In the case of clinical ultrasound, the Doppler effect will occur when there is relative motion between an echo target and the transducer, e.g. moving blood or cardiac wall. We can use this phenomenon to non-invasively detect movement and measure velocities of targets within the body.

The change in frequency between the transmitted and received sound waves at the transducer is known as the **Doppler Frequency** or **Doppler Shift.**

Imagine a little man on a red blood cell (RBC), the ultrasound target, that is moving directly towards the transducer at a velocity, v (Figure 8.2). The transducer (transmitter) is stationary and he is moving. He will see an increase in frequency $f_T' = f_T + \Delta f_T$ as he intercepts the sound waves due to his motion towards the transducer. He then re-transmits the higher frequency sound waves f_T' as an echo back to the transducer. We now have the situation of a moving transmitter (the RBC man) and a stationary receiver (the transducer).

As the RBC man moves towards the transducer, the distance between successive wave crests λ_R in his reflected wave will be shorter. A shorter wavelength λ implies a higher frequency $(c = f \cdot \lambda)$. So again, because of their relative motion, the transducer will see an increase in frequency in the echoes, on top of the already increased frequency, reflected back by the RBC man $f_R = f_T' + \Delta f_T$.

So, in the course of the ultrasound being sent out from the transducer and the echo coming back, there have been two Doppler shifts, one on the way out and one on the way back, to produce the total Doppler shift detected by the transducer $f_R = f_T + 2\Delta f_T$. This gives the factor of '2' in the Doppler Equation (shown below). If the RBC man was moving away from the transducer the Doppler effect would cause a lower frequency to be detected, again with a shift on the way out to the target and another on the return path and the sign would be negative, i.e. $f_R = f_T - 2\Delta f_T$.

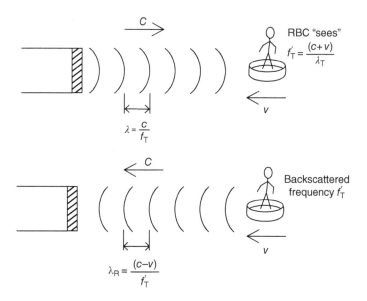

FIGURE 8.2   Illustration of the changes in frequency and wavelength seen on the outward path and the return path of ultrasound reflecting from moving blood.

THE DOPPLER EQUATION

The Doppler equation gives the change in frequency detected by a transducer when ultrasound has reflected off a moving target in the body. It assumes that the target velocity (typically, 0–10 ms^{-1}) is much smaller than the speed of sound in tissue (1540 ms^{-1}), which is true.

$$\left(f_R - f_T\right) = f_D = 2\frac{v}{c}f_T\cos\theta_D \quad so \quad v = \frac{cf_D}{2f_T}\cos\theta_D$$

where,

f_R is received frequency
f_T is transmitted frequency
f_D is the Doppler frequency or Doppler Shift
v is the target velocity
c is the speed of sound in blood

The term $\cos(\theta_D)$ is needed as only the velocity directly towards or away from the transducer generates a Doppler shift. The example of the man on a RBC assumed that he was moving directly toward or away from the transducer. In general, the target might be moving at some angle θ_D to the line of the ultrasound beam. The component of velocity toward or away from the transducer is then given by $\cos(\theta_D)$ and θ_D is known as the **Doppler angle** (Figure 8.3).

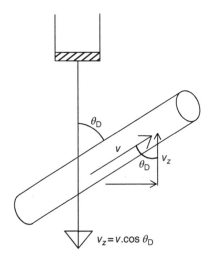

FIGURE 8.3   Definition of the Doppler angle θ_D and the component of velocity v_z toward the transducer.

NOTES

- $\cos 0° = 1$ The Doppler shift will be a maximum for a target moving directly along the line of the ultrasound beam toward or away from the transducer.
- $\cos 90° = 0$ If the target is moving perpendicular to the ultrasound beam there will be no Doppler shift.
- $\cos 60° = 0.5$ - easy to remember.
- For most echocardiography work a line of sight along the direction of motion can be obtained and so $\theta_D = 0$ and $\cos(\theta_D) = 1$. So, for cardiac work, the Doppler equation can be simplified to leave out the $\cos(\theta_D)$ term.

    The Doppler equation for cardiac work $v = \dfrac{cf_D}{2f_T}$

NOTE

Velocity v is positive when it is toward the transducer and negative when it is away from the transducer. The change in frequency is then calculated correctly.

DUPLEX ULTRASOUND

Doppler ultrasound gives us a means of making non-invasive measurements of velocity within the body. A **Duplex ultrasound** display is one which shows both the B-mode image and the Doppler Waveform. Superimposed on the B-mode display is a line showing the Doppler ultrasound beam within the field of view and the **sample**

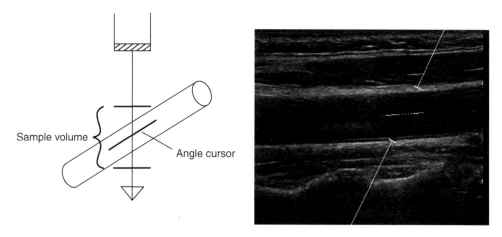

FIGURE 8.4   Diagram and image showing the alignment of the angle cursor and sample volume placement relative to the ultrasound beam and the vessel being examined.

volume from which the Doppler signal is obtained (Figure 8.4). Within the sample volume there is a rotating **angle cursor** that may be aligned with the direction of flow in the target.

When the angle cursor is correctly aligned with the direction of flow, the scanner then knows the value of θ_D. The scanner is then able to solve the Doppler equation to find the absolute velocity of the target; since it knows the transmit frequency and the speed of sound, we tell it the Doppler angle from the B-mode image, the machine compares the received frequency with the transmitted frequency to find the Doppler frequency. The target velocity v is thus found non-invasively.

EXAMPLE:

$f_T = 6\,\text{MHz}$
$c = 1540\,\text{ms}^{-1}$
$v = 0.6\,\text{ms}^{-1}$
$\theta_D = 60°$ then $\cos\theta_D = 0.5$ and $f_D = 2.4\,\text{kHz}$ which is in the audible range

---

Since f_D is in the audible range, the simplest way to detect the Doppler frequency is to amplify it and put it through a loudspeaker. This is the sound you hear when performing a Doppler examination.

CW DOPPLER

In a **continuous wave (CW)** Doppler system, ultrasound is continually being transmitted by one transducer and continuously being received by another, e.g. **hand held Doppler** used to detect peripheral pulses, or foetal heart monitor.

Figure 8.5 shows how the transmit and receive beams are arranged so that they overlap over a long thin line. The sensitive volume within which moving targets can be detected is the **cross-over region** of these two ultrasound beams when the target will be 'visible' to both beams. The cross-over region is called the **Sample Volume**.

In general, there will be a range of Doppler frequencies from many targets having different velocities within the blood stream. Higher Doppler frequencies are produced by targets moving toward the transducer and lower frequencies are produced by targets moving away from the transducer.

NOTE

When looking at blood flow, flow in the normal physiological direction is called **Forward Flow** and flow in the reverse direction is called **Reverse Flow**, e.g. in arteries forward flow is away from the heart and in veins it is towards the heart. Whether forward flow gives a higher or lower Doppler shift depends on the direction of flow with respect to the ultrasound beam because the Doppler shift depends on whether the flow is towards or away from the transducer.

Detection of Direction of Flow

In order to determine the direction of flow we need to separate out the Doppler frequencies that are less than the transmitted frequency from those that are greater. The output of the mixer detector effectively strips out the transmitted frequency and just leaves the detected Doppler frequency. We might expect the reverse flow component with lower frequencies to come out of the mixer-filter with frequencies less than zero as shown on the spectrum Figure 8.7a. However, you cannot have negative frequencies, a frequency has to have a positive value. In practice what happens is that the negative frequencies get reflected onto the positive side and forward and reverse

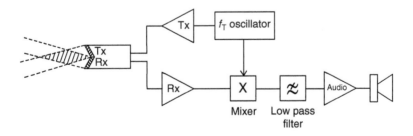

FIGURE 8.5   Block diagram of a basic hand held continuous wave Doppler system.

Detecting the Doppler Frequency

The Doppler frequency is detected by multiplying or **'mixing'** the received signal with the transmitted signal. This process gives **'sum and difference' terms**.

i.e.: sum $= f_R + f_T$ - a very high frequency
and difference $= f_R - f_T = f_D$ - the frequency we want to detect

A low pass filter allows the low frequency Doppler frequency to pass and cuts out the high frequency sum term.
    The waveforms and spectrum show this process (Figure 8.6).

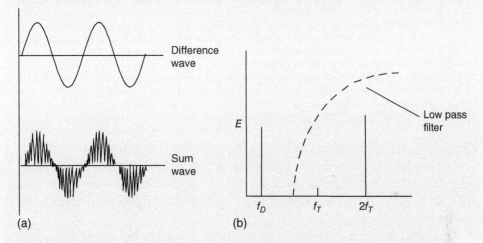

FIGURE 8.6 The sum and difference waves obtained by mixing the echo signal with the transmit frequency and the use of a low pass filter to recover the Doppler signal f_D, shown on the power spectrum.

flow frequencies overlap and cannot be distinguished. In other words, the very simple system shown in the block diagram cannot detect the direction of flow.
    In order to detect flow direction, we can mix in an **offset frequency** f_{off} into the output of our simple system to give the spectrum shown on the right of Figure 8.7. Forward and reverse flow are then separated either side of the offset frequency (which is just a few kilohertz). By filtering each one off as shown, flow direction can be detected with flow toward the probe sent to one loudspeaker and flow away from the probe sent to another, for example, left and right in a set of stereo headphones. In practice this separation is done by a more sophisticated algorithm than simply filtering the signal.
    When you hear the sound coming from the loudspeaker on the ultrasound scanner, you are hearing the actual Doppler shift frequencies being detected.

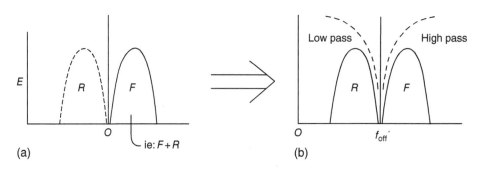

(a)   (b)

FIGURE 8.7   The principle of mixing a detected signal with an offset frequency f_{off} to separate the forward and reverse flow components. (a) Shows the situation when there is no offset frequency mixing. Forward and reversed flow signals are superimposed above zero frequency. (b) By mixing in an offset frequency forward and reverse flow are then shown either side of the offset frequency.

The higher the pitch, the higher the velocity (assuming a straight vessel) and this can be used to locate the region of fastest flow in a stenosis - if you have a good ear!

The Doppler Waveform Display (Sonogram)

The technical name for the Doppler waveform display is a **sonogram**.

We have previously seen examples of waveforms and their frequency spectra, e.g. the spectra of B-mode pulses. The mathematical algorithm that lets us convert from the time domain to the frequency domain is called the **Fourier transform** (cf. Fourier series, Chapter 5).

In digital processing, a real time signal can quickly be transformed to show its **spectrum** using a **Fast Fourier Transform (FFT)** algorithm to show what frequencies are contained within the signal.

Examples

Assume that the Doppler sample volume includes the whole cross-sectional area of a blood vessel.

1.   Slowly Moving Blood Flow in a Vessel

The **velocity profile** (change of velocities across the diameter of the vessel) has the mathematical shape of a parabola and is known as **parabolic flow** (Figure 8.8). The diagram shows the spectrum of the Doppler signal from such slowly moving blood flow. For parabolic flow, there is a constant volume of blood moving at each velocity, from zero at the vessel wall up to v_{max} in the centre of the vessel. This is reflected in the shape of the power spectrum.

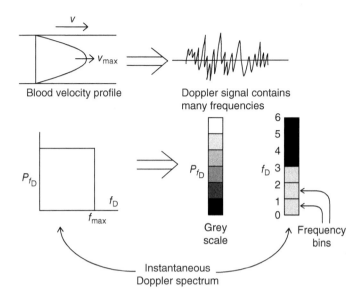

FIGURE 8.8 Diagram showing how a range of velocities gives a broad spectrum of Doppler frequencies that may be displayed using a greyscale. In this case producing the instantaneous Doppler power spectrum for slowly moving blood flow.

NOTES

- A **spectrum** is a display of the strength, or **power**, of each of the frequencies present in a signal. For example: the brightness of the colours seen in a rainbow.

- As the diagram shows the spectrum of frequencies present in the Doppler signal at one instant in time, it is called an **instantaneous Doppler power spectrum**.

- The amplitude of the Doppler signal depends on the number of red blood cells reflecting back the ultrasound signal, so the amplitude or power $P(f_0)$ of the Doppler spectrum at each frequency is proportional to the volume of blood moving at each velocity within the sample volume of the ultrasound transducer.

- The instantaneous Doppler spectrum can be shown as a graph of the power of the signal vs the Doppler frequency, or it can be shown as a vertical bar divided into a set of frequency ranges, called **frequency bins**, with each level filled in with a grey level that depends on the signal strength in that range of Doppler frequencies.

---

2. Fast Moving Blood Flow in a Vessel

The velocity profile becomes blunt and nearly all the blood moves at the same velocity, known as '**plug flow**' (Figure 8.9).

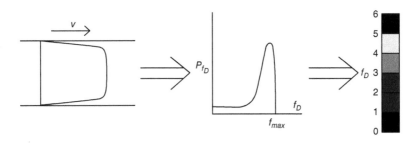

FIGURE 8.9 The instantaneous Doppler spectrum for fast flowing blood.

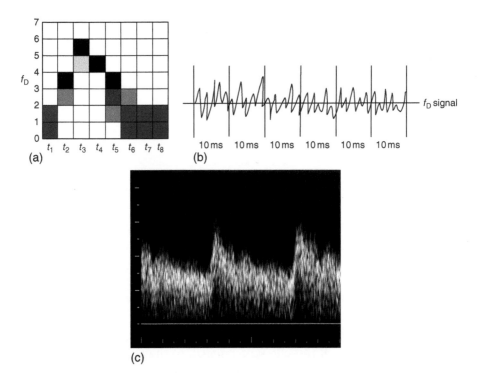

FIGURE 8.10 Diagram showing how the sonogram (a) is constructed from a series of instantaneous Doppler spectra (t_1 to t_8) each produced from consecutive 10 ms segments of the Doppler signal (b). Example of a real ultrasound sonogram (c).

As the velocity profile has changed, the corresponding instantaneous Doppler spectrum will change to show that nearly all of the blood is travelling at a high velocity, giving high Doppler frequencies. Only a small amount of blood near the vessel wall is travelling at a low velocity.

From these two examples, we see that in each case the Doppler power spectrum shows the distribution of frequencies (velocities) of the blood moving through the sample volume.

The FFT algorithm repeats the spectrum analysis of the Doppler signal very quickly to produce a series of spectra that form the Doppler waveform display or **sonogram** (Figure 8.10).

NOTES

- Each vertical line t_1, t_2.... is a complete instantaneous Doppler power spectrum.
- The length of the segment of signal used T_s determines the **frequency resolution** $\Delta f = \dfrac{1}{T_s}$ e.g. 10 ms segments gives a resolution of 100 Hz. The frequency resolution determines how fine are the steps with which we can display velocities. The significance of this is that if we want to look at low velocities with some detail, the segment length needs to be longer. If you think of looking at a very slowly moving object, you have to look at it for a long time in order to tell that it has moved.
- Each segment (e.g. 10 ms) produces 1 line of spectrum for the sonogram.
- There is a trade-off between the frequency resolution and the rate at which the number of spectra are produced. For example, 10 ms analysis segments will enable 100 spectral lines per second with a frequency resolution of 100 Hz to be displayed.
- By mixing in a base-line offset frequency, bi-directional flow can be displayed (Figure 8.11).

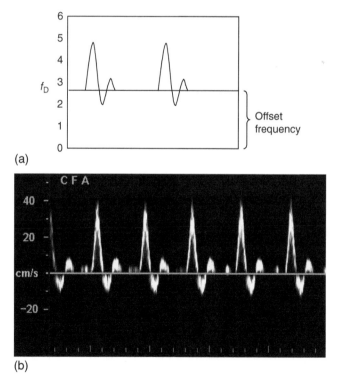

(a)

(b)

FIGURE 8.11 Showing how the use of an offset frequency enables the display of forward and reverse flow on a sonogram.

NOTE

The display can be inverted on the screen and it is common to set it so that forward flow in the physiological direction is shown above the zero line and reverse flow below the line. That is arterial flow away from the heart, when looking at arteries, and venous flow toward the heart, when looking at veins, is shown above the zero-flow line.

CW Doppler Summary

- **Disadvantage** is that any moving target in the sample volume (cross-over region) will contribute to the detected Doppler signal, that is, it has no **range discrimination**. Two blood vessels within the sample volume will both be detected and their signals superimposed on one another. Where there is only one main vessel in the sample volume, for example, in the limbs when the long arteries are being viewed, or the signal being examined is very distinct for example, when monitoring the foetal heart, this disadvantage is not too much of a problem. However, in the abdomen where there are many vessels that might fall in the sample volume, it is a big problem.

- **Advantage** is that it can detect very high velocities unambiguously. There is **no upper velocity limit** (see PW Doppler below). This advantage is useful in cardiac studies where very fast jets may be seen.

PULSED WAVE DOPPLER (PW DOPPLER) AND RANGE GATING

The way to overcome the CW disadvantage of not being able to isolate specific vessels so they can be examined individually is to use **pulsed wave Doppler** with **range gating**.

Range Gating is the method of isolating a target at a particular range. It enables specific Doppler targets to be isolated and interrogated. This is shown in Figure 8.12.

A pulse is sent out and the returning echoes are ignored until echoes from the target depth arrive. The receive 'gate' is then opened and the signal detected. Beyond the depth of interest, the 'gate' is then closed and no more signal is detected. This then defines the sensitive or **sample volume**. In other words, we have gated out the signal just from a particular range - hence range gating.

NOTES

- The **sample volume** size is a combination of the range gate length Δt and the pulse length (see below for more on this).
- The user can select the depth of the sample volume and its length. These are shown on the B-mode image (Figure 8.4).

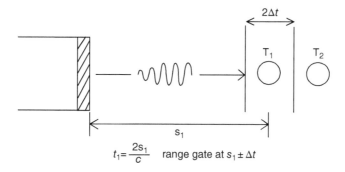

$$t_1 = \frac{2s_1}{c} \quad \text{range gate at } s_1 \pm \Delta t$$

FIGURE 8.12 Diagram showing the principle of range gating to isolate the signal from T_1 and reject the signal from T_2.

The rate at which samples are obtained is the **pulse repetition frequency** or **PRF** of the pulsed wave Doppler transmission pulses.

Detection of the Doppler Signal

The period of Doppler frequencies $(1/f_D)$ is low compared with the go-and-return time of the pulses. The length of signal gated out from each pulse is too short to determine the Doppler frequency. Therefore, the Doppler signal we want to detect has to be built up over many pulses as shown in Figure 8.13.

Tx is the sequence of transmit pulses. The range gate is opened after a delay Δt from each transmit pulse to allow echoes from the depth of interest to be detected. Each 'transmit – gated receive' sequence produces a **sample** of the Doppler signal

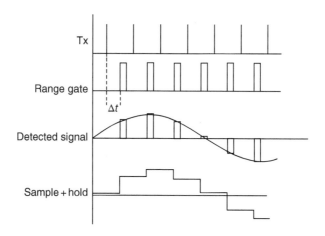

FIGURE 8.13 Diagram showing how the Doppler signal (solid line shown on the detected signal) is recovered from the range gate depth using a sequence of samples obtained from successive pulses Tx.

from the Doppler detector (mixer). The detected signal in Figure 8.13 shows the samples obtained with the true Doppler signal superimposed as a solid line.

The 'sample and hold' takes each detected level and holds it until the next value comes in. This produces the rough shape of the Doppler signal. This is then put through a low pass filter to smooth it and sent to the FFT analyser in short segments to calculate the sonogram as previously described.

> **NOTE**
>
> The key thing to note is that in PW Doppler, the Doppler signal is built up from a **sampled signal**.

A PW Doppler system is shown in Figure 8.14.

> **NOTES**
>
> - The **scale** and **sweep speed** controls affect a number of the components in the block diagram but are shown on the display for convenience.
> - In the case of **duplex scanning** with an array transducer, the Doppler beam will have **focusing** applied as for B-mode with the focus set at the depth of the sample volume.

Aliasing

In order to detect the frequency of the Doppler signal, the wave must be sampled at least twice in each period, that is, you need to know that it has at least gone above and below the zero line in that time. (recall, period $T = 1/f$).

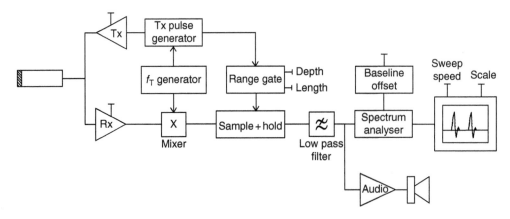

FIGURE 8.14   A block diagram of the main components of a pulsed Doppler system.

This criteria for sampling is known as the **Nyquist Limit.** The pulse repetition frequency PRF of sampling pulses must therefore be at least twice the maximum Doppler frequency we want to detect.

$$\text{i.e. PRF} = 2f_{Dmax}$$

f_{Dmax} is then the maximum Doppler frequency we can unambiguously detect with that PRF. Fewer samples than this produces a phenomenon called **aliasing** where the detected Doppler frequency gives the wrong value.

Aliasing is seen in films when a waggon wheel in a western or an aeroplane propeller appears to go backwards at certain speeds. What happens there, is that the film is made of a series of pictures (samples) shown in very quick succession so that your eye perceives continuous motion. However, if between two frames the propeller has gone all the way round but not quite, and the same again for the next frame and the next frame etc., then instead of seeing a propeller going round very fast in one direction, you will see it slowly moving backwards. It is a repeating motion that is being **inadequately sampled** (less than the Nyquist limit) and **aliasing is occurring**.

Imagine watching someone spin a ball on a string and you blink your eyes to get samples Δt apart to determine which way the ball is spinning (Figure 8.15). This is what you would see depending on how long Δt is:

(a) A high sample rate - the sample interval is much less than the period of rotation and is less than half the period of rotation. The direction of rotation can be correctly ascertained as anticlockwise.

(b) The sample rate exactly equals half the period of rotation so you can no longer tell whether the ball is travelling anticlockwise or clockwise

(c) Sample rate too low - the sample interval is greater than half the period of rotation but less than the period of rotation. The ball then appears to be going clockwise, i.e. backwards, due to inadequate or **under sampling**.

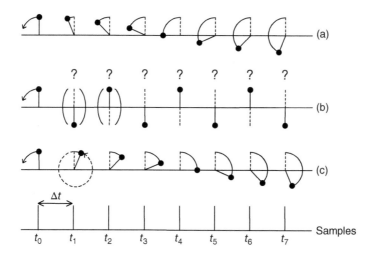

FIGURE 8.15   Rotating ball on a string showing how inadequate sampling leads to ambiguity and aliasing in determining the direction of movement of the target.

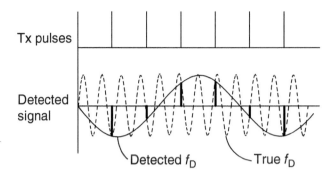

Detected f_D True f_D

FIGURE 8.16   Diagram to show how inadequate sampling of the Doppler signal leads to the detection of too low a frequency, an example of aliasing.

The same thing happens when the repeating cycles of the Doppler signal are inadequately sampled.

In Figure 8.16, the dashed line shows the true Doppler frequency. With the sample rate shown, the solid line is the frequency that is detected. Due to aliasing, instead of a high Doppler frequency f_{Dtrue}, a low frequency $f_{Daliased}$ is detected.

Appearance of Aliasing on the Sonogram

On the sonogram aliasing appears as the visual '**wrap around**' of the waveform displayed. What should be shown as a high frequency 'wraps around' to the low frequency end of the spectrum as shown in Figure 8.17.

NOTES

- Aliasing causes the high Doppler frequencies f_D to '**wrap around**' to the low frequency end of the spectrum.
- In the second diagram (b), aliasing is still seen at the low frequency end of the sonogram display, but the whole waveform has been shifted up by adding in the baseline offset frequency.
- Aliasing can be avoided by increasing the PRF to give adequate sampling - the **scale control** on the scanner increases the PRF to do this.

Maximum Velocity vs Depth

Aliasing affects the maximum velocity we can detect at a given depth in tissue:

Time for echo to return $t = \dfrac{2d}{c}$

So for a given depth d the maxPRF $= \dfrac{1}{t} = \dfrac{c}{2d}$ (need to wait for the echo to return before sending out the nextpulse)

Nyquist limit $2f_{\text{Dmax}} = \dfrac{c}{2d}$ so $f_{\text{Dmax}} = \dfrac{c}{4d}$

Doppler frequency $f_D = 2\dfrac{v}{c}f_T\cos\theta_D$, the Doppler Equation

So, maximum velocity detectable without aliasing at depth d is $v_{\max} = \dfrac{c^2}{8d\,f_T\cos\theta_D}$

The maximum velocity that can be unambiguously detected, $v_{\max}$, is reduced at greater depths as the PRF must be lower to allow for the greater go and return time for echoes. $v_{\max}$ can be increased by using a lower transmit frequency f_T. For this reason, the transmit frequency of Doppler pulses on a wideband probe is usually less than

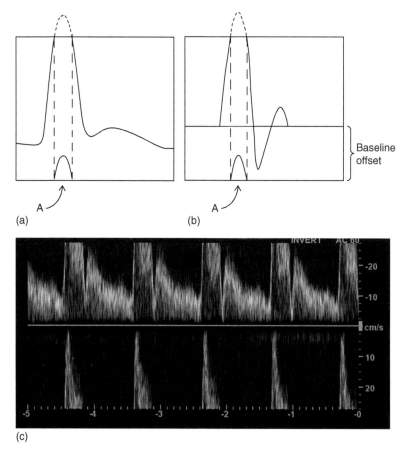

(a)   (b)

Baseline offset

(c)

FIGURE 8.17   The appearance of aliasing on the sonogram (a) showing how the high frequencies 'wrap around' to the low frequency end of the spectrum. (b) shows the case where there is a baseline offset and (c) shows an example of aliasing on a real sonogram.

the imaging frequencies used. This allows higher flows to be detected before aliasing occurs. Increasing v_{max} without aliasing may be achieved as follows:

- Adjust the PRF using the **scale control**.
- At any given depth higher velocities can be viewed without aliasing by steering the Doppler beam toward the normal to the direction of flow thereby reducing $\cos \theta$. This will give some loss in accuracy of velocity measurement (see angle error below) and at 90° there will be complete loss of signal.

Some scanners have a '**high PRF mode**'. For this they double the PRF which will double the Nyquist limit for f_D at the expense of creating two sample volumes, one at the target depth and one at half the depth (Figures 8.18 and 12.15). This is not usually a problem as there is unlikely to be another blood vessel at exactly half the distance - USER BEWARE!

For **cardiac work**, where very fast flow jets may be encountered, the normal imaging transducer may be run in CW mode, thereby avoiding aliasing altogether.

Observing Low Velocities

We have seen that in order to observe high velocities without aliasing the PRF needs to be increased by increasing the scale control. In order to observe low velocities and clearly see these on the sonogram, the PRF needs to be reduced. As discussed in relation to the sonogram above, if something is moving very slowly then you need to look at it for a longer time in order to determine that it has moved. Turning the scale down has the same effect in enabling low velocities to be seen on the Doppler waveform display.

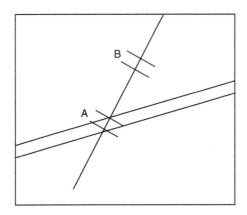

FIGURE 8.18   Doubling the PRF to double the Nyquist limit to double the Doppler frequency detectable produces a second sample volume B at the half the range gate depth A.

PW Doppler Summary

- **Disadvantage**: Subject to aliasing if the Doppler signal is inadequately sampled. There is a maximum velocity which can be unambiguously detected.
- **Advantage:** Uses range gating to enable individual targets to be isolated for examination.

PW Doppler Pulses

The pulses used for PW Doppler are generally longer than those used for B-mode imaging – typically 6–7 cycles long (Figure 8.19). This is because the transmit frequency needs to be well defined in order accurately detect changes due to moving targets. By using a longer pulse the spectrum/bandwidth of the pulse will be a lot narrower but will still not be absolutely precisely defined as shown in Figure 1.18.

The **sample volume length** will be defined as a combination of the range gate length and the pulse length (Figure 8.20). Signals will continue to be detected from along the length of the pulse as long as the range gate is open. For very short range gates, the pulse must be made shorter for the signal to be defined as arising from that range gate.

Doppler signals are noisy signals as shown in the instantaneous power spectrum, Figure 8.21.

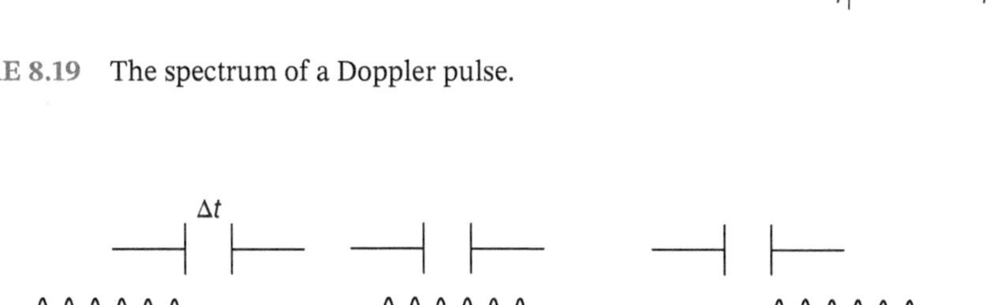

FIGURE 8.19   The spectrum of a Doppler pulse.

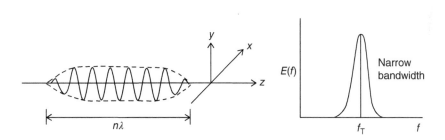

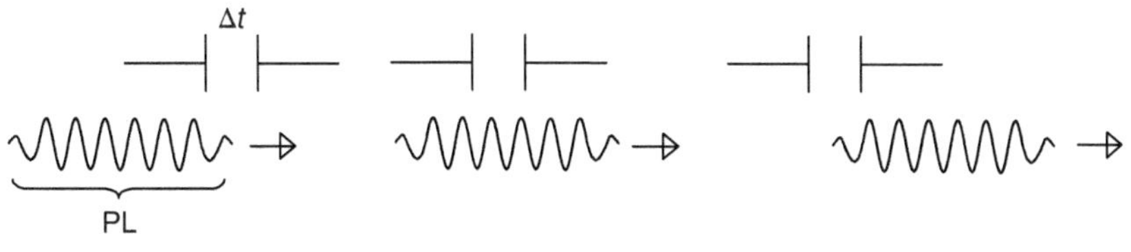

FIGURE 8.20   The relationship between pulse length PL and the range gate length Δt. Doppler signals will be detected for the whole time the pulse passes the open range gate.

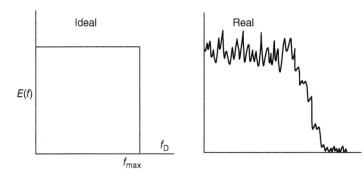

FIGURE 8.21   Diagram of the ideal and real instantaneous Doppler power spectrum for parabolic flow.

NOTES

- The received echo from the sample volume will be modulated by a speckle pattern arising from the red blood cell reflectors within the sample volume.
- Each pulse-echo sequence sees a slightly different target because the blood is moving, therefore the speckle changes with each pulse (a blood target may be considered stationary over a period of ~10–20 ms).
- As the Doppler sample volume is made shorter, for example to look at a narrow region within a vessel, the Doppler frequencies become less well defined. This is due to a phenomena called **Intrinsic Spectral Broadening (ISB)**

INTRINSIC SPECTRAL BROADENING (ISB)

In a real system, a single target moving at velocity v will produce, not just a single Doppler frequency f_D, as might be expected, but a range of frequencies $f_D \pm \Delta f_D$. This is called **intrinsic spectral broadening** or **ISB**. It results from the way scanners have to be constructed and run, that is, it is intrinsic to the real machine rather than what might be expected from an ideal machine and it cannot be avoided.

There are two causes/explanations for intrinsic spectral broadening [1].

The first cause is that all Doppler pulses will have a finite bandwidth and so produce a range of velocities across their bandwidth from a single velocity target (Figure 8.22). The shorter the pulse, the wider the bandwidth and the greater the ISB will be, i.e. $f_D \pm \Delta f_D$. A very short range gate requires a very short pulse, and hence there will be greater intrinsic spectral broadening.

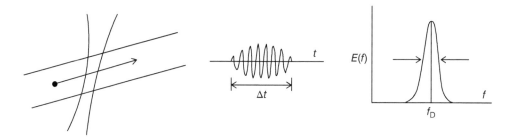

FIGURE 8.22 Diagram showing how intrinsic spectral broadening is caused by the rapid transit of a target through the Doppler beam giving a wider bandwidth of frequencies.

NOTES

- A highly focused beam (very narrow beam) will have the same effect as a very short pulse (Figure 8.22a). Any target crossing the beam will give a Doppler signal whose amplitude rapidly increases and then decreases as the target moves across the beam. Any rapidly changing signal will produce a wide spectrum in the frequency domain, as we have seen before (Chapter 1).
- Most scanners do focus the Doppler beam at the level of the range gate.
- A very short range gate will exaggerate this effect.

The second cause of ISB results from the ultrasound beam shape as shown in Figure 8.23. The beam is produced by a transducer with a finite aperture width. That means that within a single beam there will be a range of Doppler angles at any one time. A range of Doppler angles will produce a range of Doppler frequencies as given by the Doppler equation.

NOTES

- The fractional spread of Doppler frequencies is given by $\dfrac{\Delta f}{f_D} = \tan\theta_D.\Delta\varphi$

  Where $\Delta\varphi$ is the angular spread of the aperture seen from the target.
- Larger Doppler angles of insonation (i.e. closer to $\theta_D = 90°$) give greater ISB as $\tan\theta_D$ becomes large as θ_D approaches $90°$

The ISB that occurs in practice will result from a combination of these two factors. An example of ISB can be seen with a string phantom as shown in Figure 8.24. The string moves at a single constant velocity but the Doppler signal displayed will show a range of velocities due to ISB. ISB on velocity measurements made with a string phantom shows that it can be as much as ±10% about the true velocity. This will affect any peak systolic velocity measurements we make.

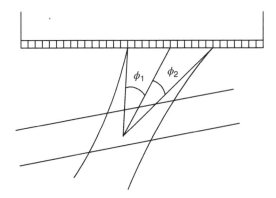

FIGURE 8.23 Diagram showing how intrinsic spectral broadening is caused by the finite width of the Doppler beam which covers a range of Doppler angles.

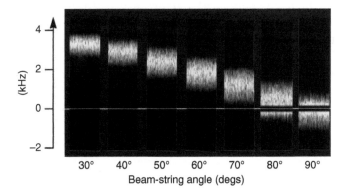

FIGURE 8.24 The spectra obtained from a string phantom showing the spread in doppler frequencies due to ISB at different angles of insonation. *Source:* Hoskins et al. [2]/with permission of Elsevier.

NOTE

In addition to intrinsic spectral broadening the Doppler spectrum may show broadening related to the real target having a broad range of velocities, for example, in the presence of turbulent flow.

QUESTION: WHAT DOPPLER ANGLE SHOULD WE USE?

Recall: The Doppler equation $v = \dfrac{c f_D}{2 f_T} \cos \theta_D$ and $\cos 0° = 1$, $\cos 90° = 0$

We frequently want to measure the **peak systolic velocity** (PSV) of a blood flow waveform. The Doppler angle that would give the highest f_D would be that arising from the edge of the beam, i.e. θ_{edge} as that has the smallest Doppler angle (Figure 8.25). So, the peak velocity seen on the sonogram is that arising from the edge of the aperture θ_{edge}.

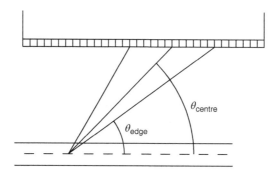

FIGURE 8.25   Definition of centre and edge Doppler angles.

However, most scanners use θ_{centre} for the Doppler angle they use to calculate the velocity. The velocity scale on the sonogram v_{scale} is proportional to $\dfrac{1}{\cos\theta_D}$ from the Doppler equation, and the velocity is therefore overestimated with a smaller value for $\cos\theta$ at θ_{centre}.

Therefore, in order to minimise ISB and the effects of centre beam calculation

USE AN ANGLE OF <60°

By using a standard, constant angle, variation in values obtained in measurements will be minimised.

USER CONTROLS

Angle Correction

The **Doppler angle cursor** needs to be aligned to the vessel/flow for the velocity scale to be correctly calibrated. Any error in positioning the angle cursor will give an error in velocity measurements (Figure 8.26).

For example, 2° error in alignment at $\theta_D = 65°$ gives an 8% error in measured velocity - therefore:

USE AN ANGLE OF <60°

Doppler Beam Steering

The Doppler beam may be steered to give a suitable angle for examining the target.

Length and Depth of Sample Volume

The target of interest is isolated by placing the sample volume (SV) over the target. The length of the sample volume should be chosen to match the examination required. In general it is good practice to set the SV to straddle the vessel. This will

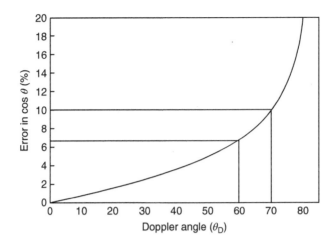

FIGURE 8.26   Graph of the percentage error in cos θ_D at different angles of Doppler angle θ_D when there is a 2° error in the measurement of θ_D.

minimise ISB and in the case of diseased vessels, ensure that the highest velocity is within the SV. This highest velocity will then always be seen as the peak velocity on the waveform.

Scale

The scale control alters the size of the Doppler waveform display within the sonogram image. It does so by changing the pulse repetition frequency of the Doppler pulses used for sampling the Doppler signal. As such, adjustment of the scale can be used to avoid **aliasing** of the Doppler waveform. In addition to being used to avoid aliasing, the scale control sets the size of the waveform display. It should be adjusted to make the waveform as large as possible without aliasing so as to ensure maximum information is visible and the most reliable measurements can be made from placing cursors on the waveform. (see Chapter 10).

Sweep Speed

The time scale across the width of the sonogram is set by adjusting the sweep speed. This should be adjusted so that the information you need for the current examination is visible and not more. For example: in most cases this might be to show 2-3 cardiac cycles of waveform. When looking for very fast changing events, such as measuring systolic rise time, the sweep speed should be increased. When looking for longer time changes, such as changes in sympathetic tone, the sweep speed should be adjusted to cover 15+ seconds (See Figure 10.13). By adjusting the sweep speed appropriately you maximise the visibility of the information you are looking for and can make the most reliable measurements.

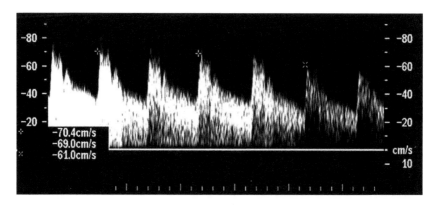

FIGURE 8.27   Sonogram obtained from the ICA with the gain being adjusted during the time of acquisition. This shows the change in PSV as the gain changes from too high at the left side to too low at the right side. The centre shows the value when there was a full range of grey levels present.

Baseline Offset and Invert

When viewing bidirectional waveforms, the baseline offset should be adjusted to show both phases of the waveform without aliasing. Inversion of the waveform may be used to ensure forward flow is positive on the sonogram.

Doppler Transmit Power

As Doppler pulses are longer than B-mode pulses and are transmitted continually along the same beam direction, the energy delivered to the insonated tissue is higher than for B-mode. The transmit power should be adjusted to as low as enables the clinical information to be obtained. Beware of over exposing sensitive tissues, e.g. the eye and early stage foetuses. This is discussed further in Chapter 11.

Doppler Gain

The gain adjusts the overall brightness of the sonogram and should be adjusted to *just* show the brightest signal as peak white. This is important as **over-gain, overestimates velocity and under-gain, underestimates velocities**. Erroneous measurements will then be made as shown in Figure 8.27.

PEAK VELOCITY ENVELOPE

Highlighting the peak velocity at each time point along the sonogram produces the peak velocity envelope (Figure 8.28). This is a useful feature to extract for characterising different **Doppler waveforms** that indicate changes in physiology or pathology in blood flow.

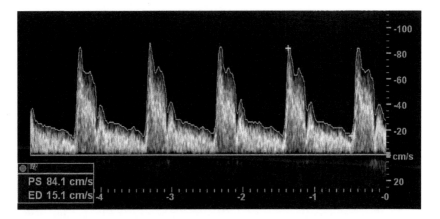

FIGURE 8.28   Example of a peak velocity envelope on a common carotid artery waveform using autotrace.

Autotrace

The scanner has a facility to automatically detect the peak velocity envelope and calculate measurements and indices directly from the peak velocity trace.

PROBLEM

How to detect the peak velocity automatically from a noisy real signal. Which point defines the maximum velocity?

Figure 8.29 shows the problem of detecting the maximum Doppler frequency. A cut-off point must be chosen. The usual way to do it is, for each instantaneous Doppler power spectrum, the scanner calculates the velocity whereby 95% (say) of the power (area under the curve) lies below the cut-off value and displays this point

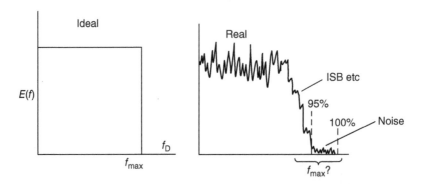

FIGURE 8.29   The spectrum of an ideal instantaneous Doppler spectrum and a noisy real spectrum showing the problem of determining where the peak value occurs.

on the sonogram as the peak velocity. This effectively discounts signal that is near the noise level and therefore more likely to cause erroneous values to be calculated.

NOTES

- Autotrace will pick up noise spikes and turbulent spikes and show them as part of the peak envelope waveform (Figure 8.30). Noise is particularly a problem when the gain is turned up for a weak signal. In making measurements, this can lead to erroneous values being displayed. The human eye-brain is very good at seeing the waveform in this noise and ignoring it - USER BEWARE!
- When the peak envelope trace is displayed on the recorded waveform it hides the true outline of the waveform making review at a later date more difficult. It is therefore good practice to record the waveform without the peak envelope as well, if the trace is used.
- If used, always ensure that the autotrace line does match the peak velocity outline of the waveform and that velocity information is not being missed or non-flow noise signal is wrongly being included.
- Using manual placement of the measurement cursor ensures you know what is being measured.

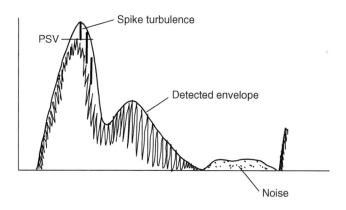

FIGURE 8.30   Showing how an autotrace waveform envelope may include spike turbulence, when the peak systolic velocity is lower as shown (PSV), and pick up noise when there is zero flow.

AVERAGE VELOCITY

The scanner calculates the average velocity from the **'normalised first moment'** of the Doppler power spectrum and may display the average velocity waveform on the sonogram. This gives the **weighted average velocity**. From this, average velocities over a period of time, e.g. 1 cardiac cycle, may then be calculated using cursors to select the time interval for calculation.

The Normalised Fist Moment

The instantaneous Doppler spectrum is divided into a series of frequency bins 1 to n, as shown in Figure 8.31. The normalised first moment is calculated by multiplying the signal power in each bin by its bin number, then summing all these results together and dividing that result by the sum of all the powers. A worked example is shown Figure 8.32.

$$\text{first moment} = \frac{\sum_1^n P(n).n}{\sum_1^n P(n)}$$

where $\sum_1^n x$ means 'sum of x from 1 to n'

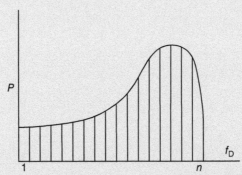

FIGURE 8.31   Illustration of how the Doppler power spectrum is divided to allow the calculation of the first moment to determine the average frequency in the spectrum.

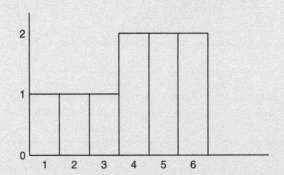

FIGURE 8.32   Example of calculating the average Doppler frequency.

Example:

$$\frac{1.1 + 2.1 + 3.1 + 4.2 + 5.2 + 6.2}{1 + 1 + 1 + 2 + 2 + 2} = \frac{36}{9} = 4 \text{ the average Doppler frequency } f_D$$

NOTES

- The average velocity will be the average from the sample volume.
- In order to get a true measure of the average velocity within a vessel, the vessel should be uniformly insonated across its whole cross-sectional area. This means setting a range gate that straddles the vessel. Unfortunately, the fact that the Doppler beam is focused at the range gate means that the beamwidth may be less than the vessel diameter and so flow near the centre-line of the vessel is emphasised as shown in Figure 8.33. As this is most often the fastest flow, the average velocity calculated will usually be overestimated.
- Set the range gate to straddle the vessel to get the nearest to uniform insonation of the vessel.
- When using calculating average velocity, different systems handle reverse flow and absence of flow in different ways, some better than others (Figure 8.34). For example, some ignore reverse flow, some average noise when there is no signal – USER BEWARE!
- Beware when using average velocity that an arterial signal is not contaminated with a venous signal and vice versa, as both will be taken into account when average velocity is estimated (Figure 8.34c).

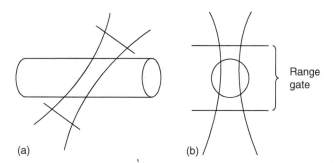

FIGURE 8.33   Placement of the range gate to straddle the vessel for average velocity measurement (a) and the problem of focusing affecting uniform insonation within the vessel (b).

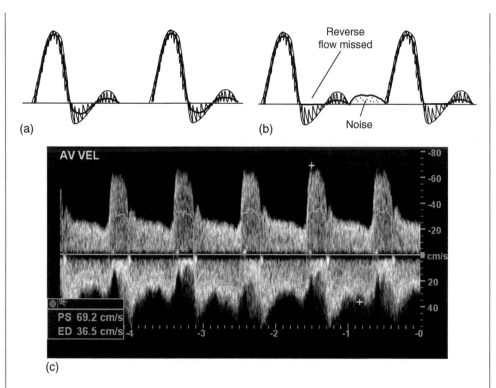

FIGURE 8.34   Showing (a) average velocity trace on the waveform and (b) an average velocity trace that erroneously does not include reverse flow and picks up noise when there is zero velocity. (c) shows an extreme example of average velocity error when an arterial signal is contaminated by a venous signal in the sample volume.

DOPPLER ARTEFACTS

As for B-mode imaging, artefacts may arise in Doppler imaging where the Doppler waveform is affected or disturbed by factors not related to the velocities of the target being imaged on the sonogram. These may affect the accuracy of measurements being made and interpretation given.

Aliasing

This artefact has been covered above.

Intrinsic Spectral Broadening

This artefact has been covered above.

Wall Thump

Blood is a very poor scatterer of ultrasound (it appears black on the image) and that is where we want to get our Doppler signal from. Arterial blood moves quickly to give high Doppler frequencies. Immediately adjacent to the flowing blood, the vessel wall is a very strong specular reflector that moves with the arterial pulse (Figure 8.35). It moves relatively slowly to give low Doppler frequencies ($f_D < 200\,\mathrm{Hz}$). The wall signal may flood the receive amplifier to give a 'thump' sound on the speakers and distort the sonogram.

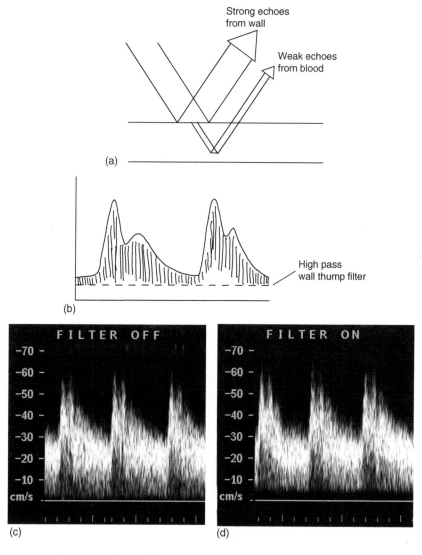

FIGURE 8.35 Illustration of the effect of poor penetration of ultrasound into a vessel (a) and the effect of a wall thump filter to remove the wall signal on the sonogram(b), (c), (d).

User Control

A high-pass '**wall thump' filter** can be applied to remove the wall signal. The level of this filter can be selected by the user. When applied, low frequencies can be seen to be missing from the sonogram as seen in Figure 8.35d.

Waveform Ghosting

In waveform ghosting, an inverted replica of the true waveform is seen below the zero line (Figure 8.36). It occurs with very superficial vessels, e.g. ankle vessels, where both forward and reverse flow are seen with in the Doppler beam aperture. Ghosting may also occur if the gain is set very high and there is cross-talk between the forward and reverse channels in the processing.

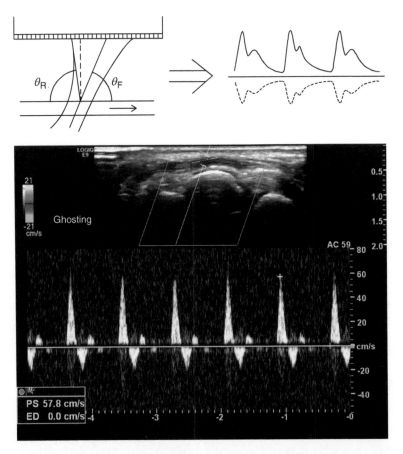

FIGURE 8.36   Showing how waveform ghosting may occur in superficial vessels when both forward and reverse signals are produced within the Doppler beamwidth.

Shadowing

The Doppler signal is obscured by dense tissues that cause shadowing on the image. This is the same as for B-mode imaging.

REFERENCES

1. Guidi, G., Licciardello, C., and Falteri, S. (2000). Intrinsic spectral broadening (ISB) in ultrasound doppler as a combination of transit time and local geometrical broadening. *Ultrasound in Medicine and Biology* 26: 853–862.
2. Hoskins, P.R., Fish, P.J., Pye, S.D. et al. (1999). Finite beam-width ray model for geometric spectral broadening. *Ultrasound in Medicine and Biology* 25: 391–404.

CHAPTER 9

Principles of Colour Doppler Ultrasound

Another way to display the Doppler shift information from moving targets within the B-mode image is to colour in the image where movement is detected. This is called **Colour Flow Mapping (CFM)** or **Colour Doppler Ultrasound (CDU)**. A map of moving targets is produced and overlaid on top of the B-mode image [1].

Problem: We cannot use the range-gate method.

We need to interrogate each gate position for at least 4–10 ms to estimate f_D (Figure 9.1).

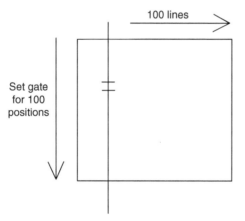

FIGURE 9.1 Diagram showing how range gating could be used to interrogate an area for Doppler information. The text describes how this would not work in practice.

Correlation

In ultrasound signal processing there are a number of situations where it is necessary to compare two signals to see how similar they are to one another or to determine the difference between them. For example, **CDU** compares the phase of RF signals from the field of view to determine where movement has occurred. **Speckle tracking** (see Chapter 14), likewise, compares segments of the detected signal to track movement of the speckle pattern in the image. Such changes may be determined using correlation algorithms.

Correlation is a mathematical procedure to compare the similarity of one signal with another. When a signal is compared with a delayed version of itself it is called **autocorrelation** and when it is compared with another signal it is called **cross-correlation**.

The procedure can be visualised as seeing one of the signals B sliding past the other signal A in a stepwise comparison to see where the greatest similarity is as shown in Figure 9.2.

B is slid past A in steps $t-2$ to $t+2$. The correlator output is at its maximum at $t=0, (t_0)$, where the two signals have the greatest similarity when compared against one another.

In the case of detecting movement in the ultrasound image, if t_0 occurs when the time difference or position between the two signals is zero, then there has been no movement between A and B.

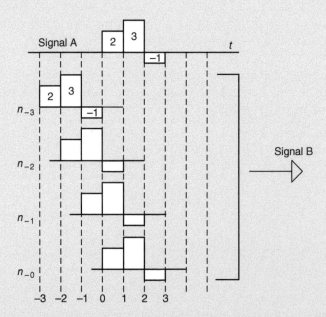

FIGURE 9.2 Diagram showing the principle of autocorrelation. A copy B, of signal A, is compared with signal A in a stepwise manner to find the point of best fit.

If, for example, the greatest similarity t_0, occurs after 5 µs or at a distance away of 0.5 mm then, knowing the timing and location of ultrasound pulses transmitted, we can calculate the speed or direction of movement needed to give those delayed similarities. Likewise, the change in phase of a signal between successive pulses can be measured.

Where the signals A and B are not exactly the same, or the signal has a lot of random noise in it, correlation can still determine where the greatest similarity is. In other words, it allows an **estimate** of the best match to be made from a noisy signal. An example is seen in the use of **coded excitation** to improve signal to noise ratio from noisy signals (see Chapter 5).

$$10\,\text{ms} \times 100 \times 100\,\text{pixels} = 100\,\text{seconds} / \text{image} - \text{not practical!}$$

Furthermore, we can only display one piece of information in each pixel, not a whole spectrum.

AUTOCORRELATION

CDU uses a method of detection called **Autocorrelation** (Figure 9.3) [2]. This compares a signal with a previous version of itself to detect what changes have taken place. It produces two output values. The mean Doppler frequency f_D and the signal power.

Looking at the speckle pattern from blood, there will be a difference in phase $\Delta\varphi$ ($\equiv$ time delay) between the two samples due to the blood cells having moved between the two pulses. Using the technique of autocorrelation, the Doppler frequency shift from the moving target can be calculated from the phase shift (see green box).

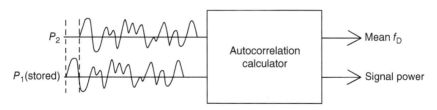

FIGURE 9.3 Showing the input and output signals from an autocorrelator.

Aside to Show that Phase Difference is Equivalent to Doppler Frequency Shift

Consider a single target moving directly away from the probe ($\theta_D = 0$).
  Movement away from probe causes phase to lag (Figure 9.4).
  Assume the phase difference between the two pulses is found to be 40°

$$\text{Let} \quad f_T = 3\,\text{MHz} \qquad c = 1540\,\text{ms}^{-1}$$
$$\lambda = 0.513\,\text{mm} \quad \text{PRF} \ 16\,\text{kHz}$$

Then with a difference in phase of 40° the pulse has travelled

$$40\,/\,360 = 0.11\,\text{of a wavelength} = 0.057\,\text{mm}$$

Between the two pulses the target moves from P_1 to P_2 (Figure 9.5).

In a time of $\dfrac{1}{\text{PRF}} = 62.5\,\mu s$ the target has moved $\dfrac{0.0567}{2} = 0.0283\,\text{mm}$ (divide by 2 for go and return time of the pulse between the two positions)

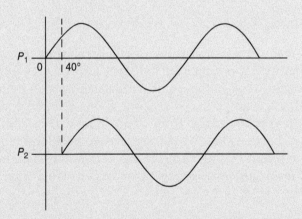

FIGURE 9.4   Example of the phase difference between two pulses P_1 and P_2 due to movement of the target between the pulses.

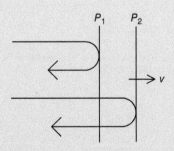

FIGURE 9.5   Showing the difference in path length for the echoes from two pulses P_1 and P_2 due to movement of the target between the pulses.

Therefore, target velocity

$$v = \frac{0.0283}{62.5 \times 10^{-6}} = 453 \, \text{mms}^{-1}$$

$$\left(\text{using velocity} = \frac{\text{distance}}{\text{time}} \right)$$

We now have the information about the movement of the target as a phase difference (40°) and as the velocity in space (453 mms⁻¹). We show the equivalence as follows:

(a) Doppler frequency using the Doppler equation

$$f_D = 2\frac{v}{c}f_T = \frac{2 \times 0.453 \times 3 \times 10^6}{1540} = 1.76 \, \text{kHz}$$

(b) Frequency from autocorrelator the phase changes by 0.11 period in 62.5 μs

$$\text{therefore time to complete 1 cycle} = \frac{62.5}{0.11} = 567 \, \mu\text{s}$$

$$\text{period } 567 \, \mu\text{s} = \frac{1}{f_D} \quad \text{so} \quad f_D = 1.76 \, \text{kHz}$$

i.e. we get the same answer from the autocorrelator as we do using the Doppler equation to calculate the Doppler frequency from the target velocity.

What the autocorrelator is actually doing is comparing the phase from the speckle pattern produced by the each of the ultrasound pulses regardless of whether it has come from moving blood or stationary tissue. As the speckle pattern is likely to change with any small movement of the probe or tissue in addition to movement due to blood flow, the output of the autocorrelator will vary significantly from pulse to pulse. This means the output signal is 'noisy' and shows random variation. The way to overcome such a 'noisy' signal is to average many pulses. Each pulse pair gives an **estimate** of the mean velocity and averaging improves the estimation. The number of pulses used to form the average is called the **ensemble length**. Scanners use a minimum of 3 pulses but more often use ~10 pulses per colour line of data. As for pulsed Doppler, the transmit pulses are several cycles long to give a narrow bandwidth with a well-defined frequency (see Figure 1.18).

To form one line for the colour map, the signal is chopped into short lengths (e.g. 1.3 μs = 1 mm length in tissue) and each segment is compared with the same segment from the previous pulse (Figure 9.6). So, pulses 1 and 2 give one sample of the Doppler frequency for each time segment, pulses 2 and 3 give a second sample etc.

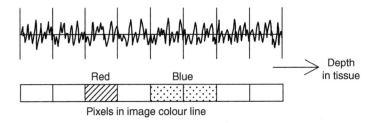

FIGURE 9.6   The output from the autocorrelator for one colour line. Each consecutive pair of returning echo pulses is broken into short segments for analysis and the output is used to colour the displayed pixels where movement is detected.

(cf. PW Doppler sampling in Figure 8.13). By this means, over the set of 10 samples collected for each CDU line, Doppler frequency estimates are obtained for a whole transmit-receive line divided into CDU pixels. This is then repeated for each colour scan line in the image.

The autocorrelator produces one value for frequency at its output for each segment comparison. In the case of a single velocity target, this will be the Doppler frequency of that target as shown in the box. Generally, there will be multiple targets within the sample volume. The autocorrelator output will then be the **mean Doppler frequency** of all the velocities present, i.e. the average phase shift.

COLOUR SCALE

The calculated mean Doppler frequencies are displayed using a **colour scale**. Change in velocity may be shown using change in hue, saturation, or luminance. A typical scale runs from dark red, through light red, to yellow for flow in one direction and dark blue, through light blue, to green in the other direction (Figure 9.7). An **offset frequency** is added to the signal to enable bidirectional flow to be shown, with flow in one direction above zero and flow in the other direction below.

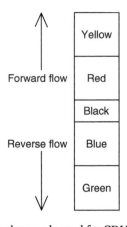

FIGURE 9.7   Example of a typical colour scale used for CDU imaging.

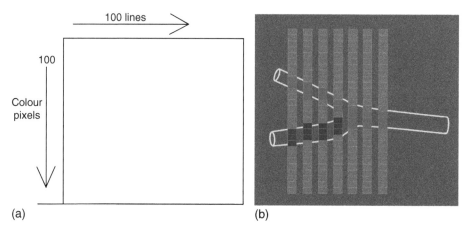

FIGURE 9.8 Diagram showing the dimensions for calculating the frame rate of a CDU image (a) and the display produced from a set of colour lines (b).

FRAME RATE

The CDU frame rate depends on the number of CDU scan lines and the depth in tissue each line interrogates, as shown in Figure 9.8.

Example: use an ensemble length of 10 pulses/line to estimate the average f_D.
100 lines across the image

$$\text{PRF for 5 cm deep scan} = \frac{1}{5 \times 13\mu s} = 16\,\text{kHz}$$

$$\text{then frame rate} \frac{16 \times 10^3 \,\text{prf}}{10\,\text{pulses}} \times \frac{1}{100\,\text{lines}} = 16\,\text{frames / second}$$

This is slow enough to look 'jumpy' if the probe is moved and the display may be cosmetically improved by averaging colour frames. Frame rate can be improved by scanning fewer colour lines and using interpolation to fill the gaps.

Because the frame rate is relatively slow it is usual to only produce a colour map over a limited area of the B-mode image. This area is called the **colour box**.

USER CONTROLS

Colour Box

The size and position of the **colour box** may be adjusted by the user to improve frame rate and image quality. As each line must be interrogated ~10 times to get a good estimate of velocity, reducing the area covered by the colour box will reduce the acquisition time significantly. Reducing the width of the colour box will have a greater effect on frame rate than reducing the depth as there are then fewer sets of 10 pulses per line to send. This is particularly important in cardiac applications where there are fast moving targets.

Frame Averaging

FRAME AVERAGING or persistence of the colour map may be adjusted by the user.

Sampling

On some machines the number of samples (ensemble length) used to form each line of the colour display may be chosen by the user. Figure 9.9 shows the trade off between increasing the number of samples to give a less noisy image but with a reduced frame rate.

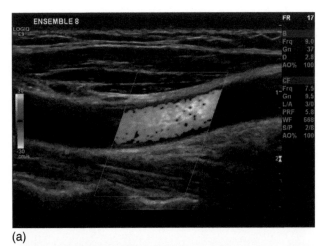

(a)

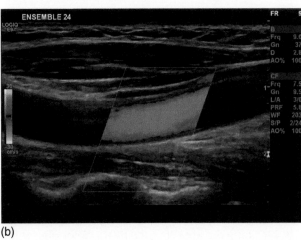

(b)

FIGURE 9.9   With frame averaging turned off, (a) shows a carotid flow with an ensemble length of 8 and a fame rate of 17 fps, (b) shows an ensemble length of 24 and a frame rate of 9 fps.

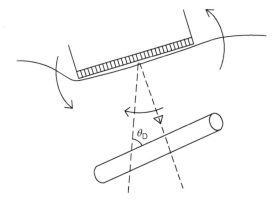

FIGURE 9.10   Illustration of how to 'heel and toe' the probe to improve the Doppler angle.

CDU AND THE DOPPLER ANGLE

As for CW or PW Doppler, the CDU signal is governed by the Doppler equation with its $\cos \theta$ term, and hence there needs to be an angle between the ultrasound beams used to form the colour image and the blood flow being examined. This may be achieved in one of two ways. The colour box may be **steered** left or right relative to the orientation of the B-mode image. CDU ultrasound beams are then electronically steered as shown by the box. This helps to give an acceptable Doppler angle to the blood flow seen.

The Doppler angle to the flow may be further improved by '**heel and toeing**' the probe as shown in Figure 9.10.

As $\cos 90° = 0$, there is no colour signal when flow is at 90° to the CDU ultrasound beams and blood will appear black as on the normal B-mode image.

IMPORTANT NOTE

The colour value displayed will change if either the **velocity changes** or the **Doppler angle** changes. For example. In the case of a tortuous vessel the Doppler angle changes along its length, so even if the flow velocity is constant along its length, the colour shown on CDU will change.

In the illustration shown (Figure 9.11), the flow changes direction relative to the ultrasound probe. When it is at 90° to the probe there is no flow detected ($\cos 90° = 0$) and the colour is black.

COLOUR ALIASING

The colour Doppler signal is built up from a series of pulse pairs that are compared in the correlator to give an estimate Doppler frequency. As with PW Doppler, these pulses pairs are subject to the **Nyquist limit**. That is, they must be less than half a

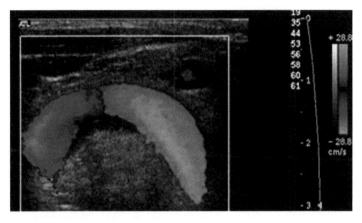

FIGURE 9.11   A CDU image of flow in a curved vessel.

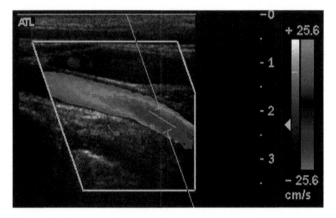

FIGURE 9.12   Illustration of colour aliasing in a uniform vessel due to change of vessel angle with respect to the ultrasound beam. High Doppler frequencies are erroneously shown at the low end of the colour Doppler spectrum.

period of the maximum Doppler frequency apart, i.e. >2 samples per cycle, in order to unambiguously estimate the Doppler frequency. If the time between samples is longer than this the colour display will show **aliasing**. When aliasing occurs the high Doppler frequencies will appear as low frequencies. On the colour map this appears as a '**wrap around**' from one end of the colour scale to the other, in a similar way to the sonogram waveform showing 'wrap around' with aliasing. (see Chapter 8).

The image in Figure 9.12 shows the colour going from yellow to green in the centre of the vessel, i.e. 'wrap around' due to aliasing, where there are Doppler frequencies that are being inadequately sampled. To avoid aliasing the **scale** needs to be increased, which increases the pulse repetition frequency to bring it above the Nyquist limit.

The way to remember the colour changes seen is '**Over the top – high velocities and aliasing, red through black to blue – change in direction of flow**' (Figure 9.13)

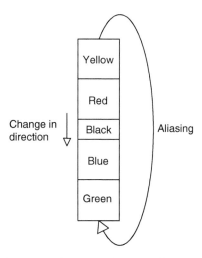

FIGURE 9.13 Showing the distinction between aliasing and a change in direction of flow.

USER CONTROLS

Scale

- The scale control alters the pulse repetition frequency of the pulses used to form the colour lines. It may be used to avoid **aliasing** by increasing the PRF above the **Nyquist limit**.
- It scales the detected velocities against the colour scale. In other words, it can be adjusted so that the velocities present in the flow are mapped so as to cover the whole colour range of the colour scale.
- Where it is required to detect fast flow, e.g. in a jet caused by disease, the scale may be set so that for normal flow the peak colour is just below aliasing. Then, when flow velocities exceed that level aliasing will occur and will easily be seen on the CDU display, highlighting where fast flow is as seen in Figure 9.14.

IMPORTANT NOTE

Unlike duplex imaging, where the Doppler angle cursor is used to tell the scanner what the Doppler angle is, in CDU the angle is unknown. Blood vessels and flow being imaged may be at any angle to the ultrasound colour beams. The colour scale shown on the display may have a number shown against the highest colour that changes as the scale control is altered. This number is an *indicative number only* for the scale setting. It shows the peak velocity for flow that is parallel to the colour lines and should *not* be used as a velocity measure.

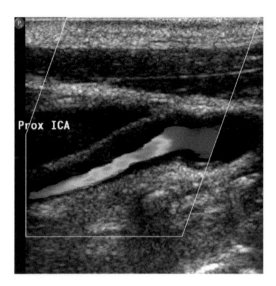

FIGURE 9.14   Aliasing in a straight vessel showing where a stenosis is causing increased blood velocities.

Invert

- The colour scale may be inverted, for example, to show flow towards the probe as red and flow away as blue, or alternatively, to show arterial flow as red and venous flow as blue. However, in **echocardiography** it is conventional to always use red towards the probe and blue away from the probe.
- The colour scale on the scanner display will always show flow toward the probe above the zero line. This may be red or blue depending on the invert setting. **The user should always be aware which way the colour is set up**, especially when looking for pathological **reverse flow**.
- When steering the colour box to the left and the right, the colour scale may or may not be inverted by the scanner - USER BEWARE!
- If you are holding the probe the wrong way round in your hand the colours will be reversed from what you expect.

When looking for **reverse flow** (retrograde flow) always check:

- Is the probe the right way round in my hand?
- With the box steered, is the flow toward the probe, as shown on the displayed scale (red or blue), in the physiological direction expected?
- Only then can you confidently diagnose reverse flow.

DISCRIMINATION OF STATIONARY TARGETS

Ideally, a vessel with moving blood should have colour showing up to the wall of the vessel, then the wall and surrounding tissue should be shown in grey scale on the B-mode image. However, as with PW Doppler, a high pass **wall filter** must be used to cut out the clutter signal from adjacent tissue that moves due to the cardiac pulse and would otherwise give a CDU signal outside the vessel. Even with a wall filter, discrimination of surrounding tissue is not perfect. This leads to **drop-out**, where there is no colour filling in the vessel adjacent to the wall, and **colour bleeding**, where colour bleeds over the edge of the vessel lumen into the wall and surrounding tissue (Figure 9.15).

Imaging Small Vessels

The use of a wall filter also has an effect on the visualisation of small vessels. CDU is not reliably able to visualise vessels <1 mm diameter. This is because:

- Their velocities are low and produce Doppler shifts below the wall filter cut-off frequency.
- Small vessels themselves move at the same rate as their surrounding tissue.

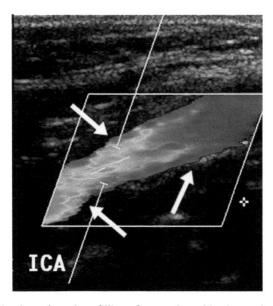

FIGURE 9.15   showing how the colour filling of a vessel can bleed over the vessel wall or not fill the lumen.

More recently viewing these very small vessels and vessels with slow moving blood has become possible with ultrafast ultrasound and is described in Chapter 13.

USER CONTROLS

Wall Filter

As with PW Doppler, a high pass filter can be applied to the CDU signal to help with low velocity wall movement. The cut-off frequency of this filter is adjustable by the user.

B-Mode Priority

The colour is overlaid on the B-mode image. Related to the question of discrimination of movement is whether the display gives preference to showing B-mode or colour. The user can select the priority given to colour based on the level of grey displayed in the B-mode image. This is often shown as a marker at the side of the grey scale wedge on the display. If this is set too low on the grey scale there will be drop-out of the colour signal as a low grey level is given priority as shown in Figure 9.16.

POWER DOPPLER (PD)

There is a second mode of displaying the colour information on the B-mode image called **power Doppler**. Instead of using the directional velocity information from the autocorrelator, the signal strength or spectral power of the Doppler signal is displayed as a colour map regardless of frequency or direction (Figure 9.17). This is proportional to the number of moving reflectors detected (effectively, red blood cells).

NOTES

- Because the velocity does not have to be estimated from a noisy signal, the display of power Doppler appears smoother and is less dependent on Doppler angle, so there is better colour filling at low velocities and close to 90°.
- It has higher sensitivity than directional CDU.
- Power Doppler is often shown using a blue–orange flame-type scale.
- Useful for viewing:
  - Tortuous vessels where CDU would change colour as the vessel changes direction. Power Doppler shows a uniform colour.
  - An arterial tree, e.g. kidney or vessel malformation – vessels in many directions show up in a uniform colour (Figure 9.18).

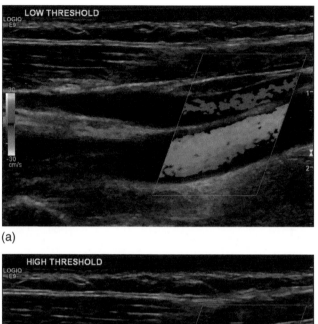

(a)

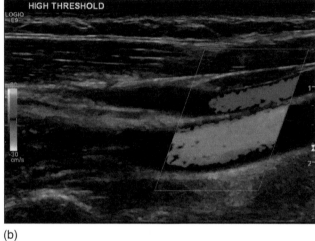

(b)

FIGURE 9.16 Showing the effect of B-mode priority (a) with a low threshold for B-mode display and (b) a high threshold. The difference is seen more clearly in the vein shown in blue.

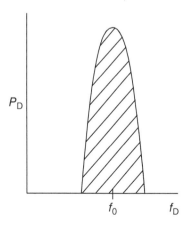

FIGURE 9.17 Power Doppler shows the strength of the Doppler signal which is equivalent to the area under the curve of the Doppler spectrum.

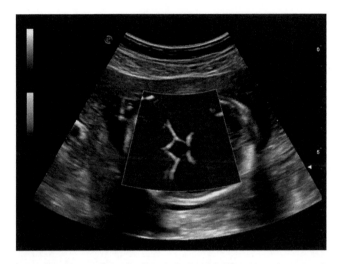

FIGURE 9.18   Power Doppler image of a foetal circle of Willis. Note that vessels with flow in different directions are shown uniformly. *Source:* Courtesy: Esaote.

NOTE

Some machines improve the quality of the directional colour display by using the power Doppler information and colouring it in with the directional information.

A variation on the conventional way to display CDU and PD images is to use the velocity (CDU) or power (PD) scale to give a '3D' appearance to the displayed image as shown in Figure 9.19. The higher values of the colour velocity are used to brighten the displayed image. By enhancing the display of flow in this way the vessel gains a 3D effect as the fastest flow tends to be along the centre line of the vessel with slow flow adjacent to the vessel wall. This is seen in Figure 9.19b.

CDU ARTEFACTS

As for B-mode artefacts, some CDU artefacts can be useful.

Aliasing

This can be a useful artefact as it can show where velocity increases at a stenosis in a vessel - hence it can show where the PW Doppler sample volume needs to be placed to measure the highest velocity. See Figure 9.14.

Shadowing

CDU is subject to shadowing from dense/reflecting structures in tissue causing **colour dropout** (Figure 9.20)

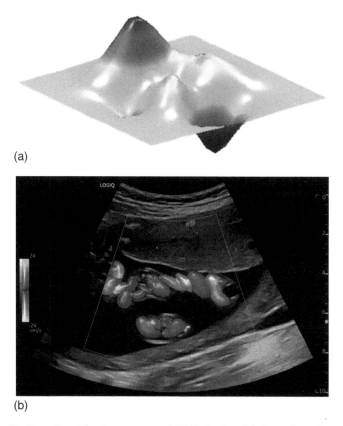

(a)

(b)

FIGURE 9.19 'Radiant flow™' enhancement of CDU display. (a) shows how the colour scale is used to give a '3D' effect to the displayed CDU map. (b) shows an example of the umbilical vessels. *Source:* GE.

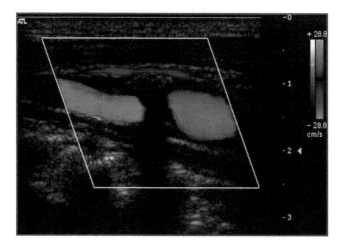

FIGURE 9.20 Carotid artery showing dropout in the colour distal to a shadow casting plaque.

Wall Discrimination

> **NOTE**
>
> Because of dropout and colour bleeding near vessel walls, colour should *not* be used to make diameter measurements of vessels. Instead, use a good quality B-mode image where the walls are clearly imaged.

Flash Artefact

This is caused by probe movement on the skin causing colour signal to be generated from the tissue moving relative to the probe. It appears as colour flashes across the whole image. Algorithms to suppress flash artefact mean that it is much improved in modern scanners.

Mirror Image Artefact

As for B-mode, colour can be shown where the beam has reflected from a surface in the body, e.g. from bone at the ankle for the tibial vessels, from the lung for subclavian artery (Figure 9.21).

Twinkle Artefact

Calcified vessels and stones can show a colour 'twinkle' artefact that is not related to blood movement (Figure 9.22). Its origin is not clearly understood but may be similar to comet tail artefact in B-mode, i.e. reverberations generating a spurious Doppler signal. It appears as colour sparkling at points behind the calcification. Its occurrence is sensitive to machine settings [3].

Temporal Artefact

The frame rate for CDU is relatively low and at 20 fps it takes 50 ms to obtain 1 frame. During the systolic rise time arterial flow velocity can increase by $1\,\text{ms}^{-1}$ in 90 ms. In other words, in the time it takes to obtain one frame, the flow can significantly change. This gives rise to a temporal artefact whereby one side of the colour box can show a different colour than the other, even though it is one uniform straight vessel being viewed (Figure 9.23).

COLOUR SENSITIVITY

It can be difficult to get colour filling in a deep vessel when there is poor visualisation or flows are very low. CDU needs a Doppler angle. This can be achieved either by beam steering or by 'heel and toeing' the probe. Recalling from Chapter 3 that

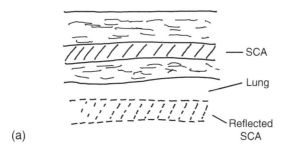

(a)

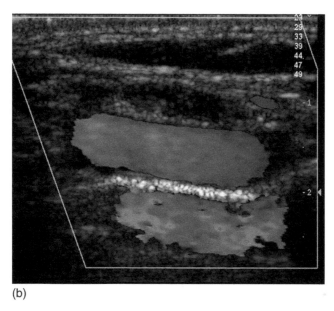

(b)

FIGURE 9.21 Reflection of the subclavian artery (SCA) seen within the lung after reflection from the top surface of lung.

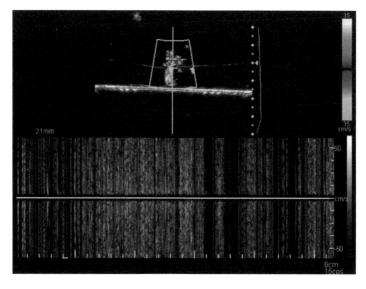

FIGURE 9.22 An example of twinkle artefact from a stone in water showing that its sonogram is just noise. *Source:* Hirsch et al. [4]/Permanyer.

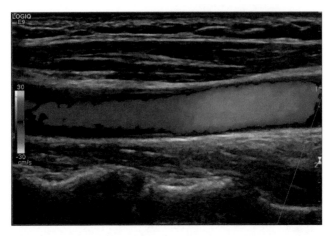

FIGURE 9.23   Temporal artefact showing a change in velocity along a common carotid artery due to the time taken to acquire the image. Frame rate was reduced to 3 fps to clearly show the effect.

the transducer has its greatest sensitivity when the beam is straight ahead with no steering, sensitivity is improved by steering straight ahead and then using 'heel and toe' to give a Doppler angle to the vessel.

For low flows, turn the colour scale down and turn the gain up until colour noise is *just* apparent to achieve greatest sensitivity.

PRESETS

Manufacturer presets for different examinations will alter the CDU settings discussed. As the user, it is important to make sure the settings are appropriate for the patient in front of you and that you re-adjust the settings as necessary to achieve optimal CDU imaging.

COLOUR M-MODE

M-mode enables tissue movement along a single line of sight to be appreciated over a period of time (Chapter 5). Using colour Doppler on an M-mode display enables changes in blood velocity and direction along that line to be visualised. As only one scan line is being imaged there is high temporal resolution and the technique is useful within echocardiography (Figure 9.24).

TISSUE DOPPLER IMAGING (TDI)

Tissue Doppler imaging is used in cardiac examinations. It is based on the principle that blood moves quickly whilst the myocardium moves relatively slowly.

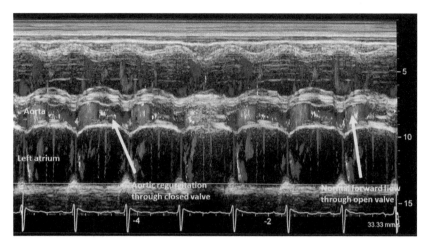

FIGURE 9.24 A colour M-mode display of flow across an aortic valve with regurgitation.
Source: Courtesy: Chris Eggett.

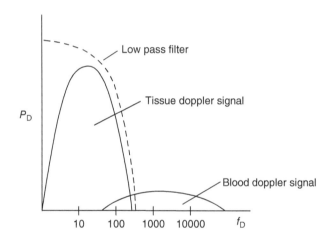

FIGURE 9.25 Spectrum showing the low pass filter used to isolate the tissue Doppler signal.

Instead of using a high-pass filter to filter out the wall movement, the reverse is done (Figure 9.25). A low-pass filter is used to filter out the blood signal and allow the cardiac wall movement to be shown in colour (Figure 9.26). This enables areas of myocardial ischaemia to be identified as the affected walls move paradoxically to normal myocardium.

TDI will detect any tissue motion, both passive, where there is inactive myocardium or due to tethering of tissue to a moving region, and active motion due to muscle contraction.

If all points in a region of the image move together, we have **displacement** of the target. If different points move at different velocities, we have **distortion** or **deformation** of the target, and the target is changing shape (Figure 9.27).

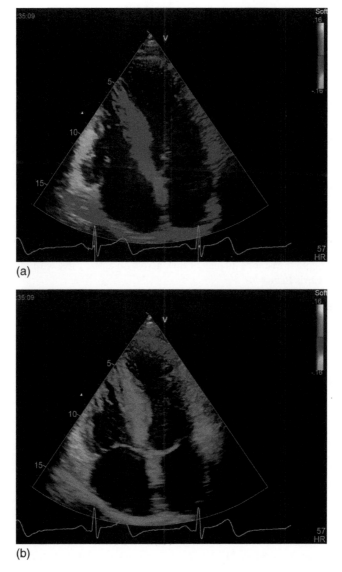

FIGURE 9.26   A tissue Doppler image of a 4 chamber view showing movement in the normal heart in (a) diastole and (b) systole. *Source:* Courtesy of Chris Eggett.

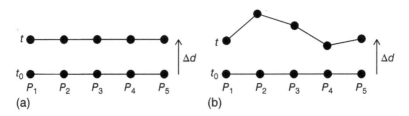

FIGURE 9.27   Showing position of points in a structure P_1 to P_5 at times t_0 and t. (a) All points have moved the same distance Δd and we have displacement of the structure. (b) The points have moved with different velocities and their positions at t are at varying distances and we have distortion or deformation of the structure.

MYOCARDIAL STRAIN IMAGING

A further development from simple observation of cardiac motion with TDI is to measure myocardial strain and strain rate. This can give detailed information on how the myocardium is performing and differentiate active versus passive movement [5].

Strain

Strain ε is the increase or decrease in length of a dimension of the myocardium over a reference length L_0, for example, the length at end diastole in the cardiac cycle, in response to an applied force (Figure 9.28). This is also called the **Lagrangian strain**.

$$\varepsilon = \frac{L_t - L_0}{L_0} \times 100$$

It is a dimensionless number expressed as a percentage, positive for an increase in length and negative for a decrease in length, where L_t is the length at time t.

Strain Rate (SR)

Strain Rate (SR) is the rate of change of strain, that is how quickly the change in length is taking place.

$$SR = \frac{\Delta \varepsilon}{\Delta t} = \frac{\Delta L}{L_0} \cdot \frac{1}{\Delta t} = \frac{\Delta v}{L_0} \text{ equals the time gradient of } \varepsilon \left(\text{unit} : s^{-1} \right)$$

SR is positive on lengthening and negative on thinning or shortening. It is a measure of the velocity of wall movement from point to point.

Three dimensions of strain are considered as shown in Figure 9.29:

1. Longitudinal strain L (base – apex of ventricle)
2. Radial strain R (thickening of myocardium)
3. Circumferential strain C (change in length as the circumferential line contracts and relaxes)

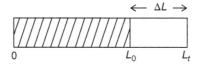

FIGURE 9.28   The dimensions for measuring strain ε.

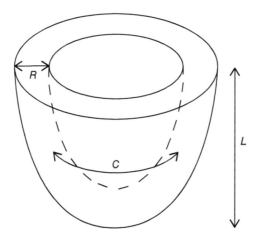

FIGURE 9.29   The three dimensions of strain within a cardiac chamber considered in cardiac examinations.

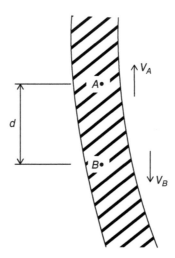

FIGURE 9.30   Points A and B selected to measure the velocity difference using TDI for calculating SR.

Using TDI to Measure Strain

Two regions of interest (ROI), A and B a distance d apart, are selected on the image (Figure 9.30). Using TDI, the Doppler velocity is measured at points A and B to obtain the velocities v_A and v_B at these points. The difference in velocities is the velocity gradient Δv from which the SR can be calculated:

$$\frac{\Delta v}{d} = \frac{v_B - v_A}{d} = \text{SR}$$

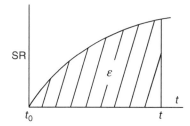

FIGURE 9.31   Strain ε as the summation of change in strain rate over time.

The strain may then be calculated from the summation of the strain rate over time:

$$\varepsilon = \int_{t_0}^{t} SR.dt$$

i.e. the area under the SR curve over time as shown in Figure 9.31.

Using TDI to calculate strain and strain rate has a number of limitations.

- It is dependent on the Doppler angle of incidence between the ultrasound beam and the myocardial motion. If this is >20° the strain and SR will be significantly underestimated.
- It requires an acquisition rate of >130 frames per second.
- It needs a good B-mode image with optimal wall-TDI alignment.
- The Doppler scale must be adjusted to avoid aliasing.
- There is a trade-off between the size of the ROI for A and B which determines the SNR and spatial resolution.

TDI derived strain and SR has a typical variability of 10–15%.

These limitations of TDI measured strain and SR are largely overcome by the use of **speckle tracking** from the B-mode image to measure strain.

SPECKLE TRACKING ECHOCARDIOGRAPHY (STE)

A series of 2-D B-mode images from across the cardiac cycle are stored (typically, 3 cycles are obtained) and the movement of the speckle pattern is tracked across the images using a technique such as **cross-correlation** between the images (Figure 9.32). Movement may be detected in both lateral and axial directions to enable speckle paths to be identified and velocity vectors for actual direction of movement to be calculated [6] (see Chapters 13 and 14 for descriptions of vectors and speckle tracking)

From the velocity vectors, strain and SR may be calculated along the vector as for TDI velocities described above. Analysis is performed on a segment by segment basis as well as global longitudinal values.

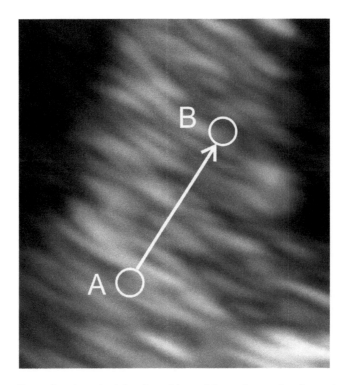

FIGURE 9.32   Illustrating the principle of speckle tracking to locate the change in position of speckle 'markers' in the B-mode image.

Torsional motion (twisting) may be calculated by measuring rotations (circumferential motion) at two levels. Torsion is then the net difference between the two rotations in degrees/second.

STE has a number of advantages over TDI strain imaging.

Pros:
- Velocities are not dependent on Doppler angle.
- Deformation may be measured in more than one direction.
- Higher spatial and temporal resolution than TDI.
- Optimal frame rate for B-mode acquisition is 50–70 fps.
- Better reproducibility than TDI method.
- Fewer artefacts seen than with TDI.
- Less sensitive to signal noise.

Cons:
- Needs high quality B-mode images which may limit use in some patients.
- Signal may be lost with translational motion of the heart out of the scan plane. This may be avoided where 3-D tracking data is available.

- Processing is off-line and requires manual input to identify and outline the endocardial border on one image in order for analysis of the whole image set to be performed. Newer techniques using advanced algorithms are able to overcome these limitations as discussed in Chapter 12.
- Because different algorithms are used by different vendors, measurements should not be compared across different manufacturers scanners.

STE requires good quality B-mode images with optimal setting of gain, focus and image width and depth centred on the cardiac chamber of interest. Breath holding by the patient will minimise extraneous movement which can create artefacts. In particular, the endocardial and epicardial borders should be clearly delineated. A good quality ECG signal is also required for gating.

STE Display

The velocity, strain and strain rate data may be displayed in a number of ways. They may be shown graphically as a change over the cardiac cycle, as a colour mapping over the B-mode image or as a segmental display known as a **Bull's Eye Plot** that displays peak systolic strain in each segment of the myocardium (Figures 9.33 and 9.34).

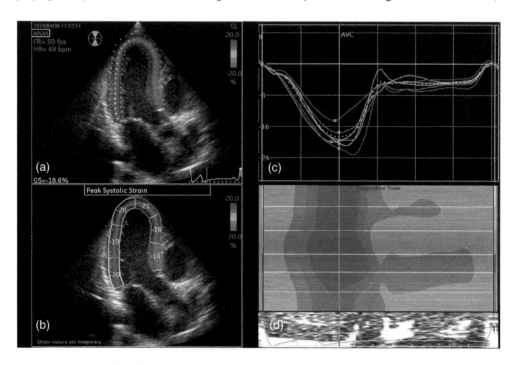

FIGURE 9.33 Regional strain around the normal left ventricle. Panel (a) shows the points from which the strain has been calculated. Panel (b) shows the regional strains at peak systole with each region colour coded. Panel (c) shows the strain for each region over the cardiac cycle, and panel (d) shows a heat map of the strain over time for the measured points going round each region versus time on the x-axis. *Source:* Courtesy of Chris Eggett.

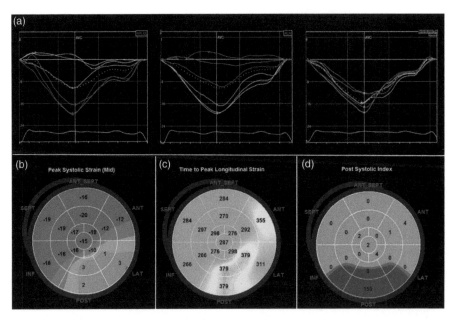

FIGURE 9.34   Segmental strain traces recorded from apical 3-, 4-, and 2-chamber views, respectively (a); bull's eye plots displaying midwall peak systolic strain (b); time to peak longitudinal strain (c); and post-systolic index (percentage of post-systolic increment over systolic strain) (d) in a patient with a left circumflex infarct. *Source:* Collier et al. [8]/with permission of Elsevier.

REFERENCES

1. Evans, D.H., Jensen, J.A., and Nielsen, M.B. (2011). Ultrasonic colour Doppler imaging. *Interface Focus 1*: 490–502. https://doi.org/10.1098/rsfs.2011.0017.

2. Hedrick, W.R. and Hykes, D.L. (1995). Autocorrelation detection in color Doppler imaging: a review. *Journal of Diagnostic Medical Sonography* 11: 16–22.

3. Kamaya, A., Tuthill, T., and Rubin, J.M. (2003). Twinkling artifact on color Doppler sonography: dependence on machine parameters and underlying cause. *American Journal of Roentgenology* 180: 215–222.

4. Hirsch, M., Palavecino, T., and León, B. (2011). Color Doppler twinkling artifact: a misunderstood yet useful sign. *Revista Chilena de Radiología* 17: 82–84.

5. Nesbitt, G.C., Mankad, S., and Oh, J.K. (2009). Strain imaging in echocardiography: methods and clinical applications. *The International Journal of Cardiovascular Imaging* 25: 9–22. https://doi.org/10.1007/s10554-008-9414-1.

6. Johnson, C., Kuyt, K., Oxborough, D. et al. (2019). Practical tips and tricks in measuring strain, strain rate and twist for the left and right ventricles. *Echo Research and Practice* 6: R87–R98. https://doi.org/10.1530/ERP-19-0020.

7. Blessberger, H. and Binder, T. (2010). Non-invasive imaging: two dimensional speckle tracking echocardiography: basic principles. *Heart* 96: 716–722.

8. Collier, P., Phelan, D., and Klein, A. (2017). A test in context: myocardial strain measured by speckle-tracking echocardiography. *Journal of the American College of Cardiology* 69: 1043–1056. https://doi.org/10.1016/j.jacc.2016.12.012.

CHAPTER 10

Making Measurements

A great deal of the usefulness of ultrasound imaging arises from measurements that can be made from the images. Values obtained may be compared to classify normal from abnormal and to monitor changes over time. It is important, therefore, to understand the limitations on measurements made and to consider the errors that may arise. There are a number of key concepts relating to measurement which we consider followed by a consideration of how they apply to making measurements on ultrasound images and Doppler waveforms. We begin with a discussion on accuracy and precision (Figure 10.1).

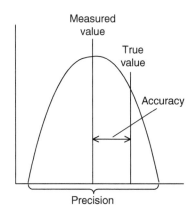

FIGURE 10.1 Graphical display of a range of measurements showing the accuracy and precision of a measured quantity.

Ultrasound Technology for Clinical Practitioners, First Edition. Crispian Oates.
© 2023 John Wiley & Sons Ltd. Published 2023 by John Wiley & Sons Ltd.

ACCURACY

The accuracy of a measurement is its closeness to the true value.

EXAMPLE 1

If you ask a friend the time and they look at their watch and say my watch says 10.00am and then immediately you see a radio clock tied to atomic time and see that it is in fact 10.05am, the time they gave you is inaccurate by 5 minutes. That may or may not matter to you, depending on what you need to know the time for. The time they gave is right to within 5 minutes. The radio clock, on the other hand, is always telling the true time as it is tied to an atomic clock from which the passage of time is defined.

EXAMPLE 2

If you had a homemade ruler where you had drawn the marks on the ruler by hand and made each centimetre gap approximately 1 cm but on average you had made them 2 mm too large, then any measurements you made with the ruler would be inaccurate and the error would be larger for larger distances. The error for any one measurement would vary in a random way but the average error would be 20% too large.

PRECISION

Precision is the range of variation of measured values around the average measured value.

EXAMPLE 3

If you asked your friend the time at several points during the week and every time the answer they gave you was exactly 5 minutes behind the radio clock, then their watch could be considered precise, though inaccurate. Although, once you knew there was a 5-minute offset, you could add 5 minutes and obtain the accurate time. This may be compared with the situation where it was 5 minutes behind, then 3 minutes behind then 6 minutes behind etc. but on average was 5 minutes slow. In this case the precision is ±2 minutes (say).

Another way of looking at precision is to ask how closely can I specify the value I measure to the actual value I believe I have measured – which I assume to be the true value.

> ## EXAMPLE 4
>
> If you have a ruler that you believe to be accurate – the distance marks are spot on – but which is only marked in centimetres, then you can estimate the gap between the marks to maybe a quarter of a centimetre. However, you would not be justified in quoting a value to 3 decimal places as you cannot be that precise in your measurement. The precision of a measurement will limit how finely the value can be meaningfully specified.
>
> Accuracy and precision are closely connected but are different concepts. The atomic clock is both accurate and precise (accurate to 1 second in millions of years and precise to 1 part in 10^{10}).

Example 1 is precise but inaccurate, Example 2 is inaccurate and imprecise and Example 4 is accurate but imprecise.

These examples illustrate that precision may be limited by the finesse of my measuring device or be limited by random events occurring in the measuring process.

HOW ACCURATE OR PRECISE DO WE NEED TO BE?

Depends on:

- What you want to use the data for, for example, comparing kidney size or dating a pregnancy require different levels of precision.
- What you are comparing it with, e.g. CT, MRI, measuring tumour growth from a previous examination, standard normal values.
- What the normal range of variation is – a large normal range does not require a highly accurate measurement to confirm normality.

Accuracy and precision become crucial when you are pushing the limits of what ultrasound imaging is capable of resolving or when small errors make a significant difference in diagnosis and clinical decision making. For example, measuring nuchal translucency, or intimal thickness in arteries.

REPRODUCIBILITY

For the purposes of making measurements in a clinical setting, a closely related concept is that of reproducibility. If I make a measurement, do I get the same value when I repeat that measurement – **intraobserver variability** – or when someone else makes the same measurement– **interobserver variability**?

Reproducibility is likely to be poorer when the same target is measured on different scanners and caution needs to be exercised when comparing measurements made across scanners. Such measurements require the scanners to be calibrated against standard test objects to ensure accuracy and precision are within acceptable limits.

SYSTEMATIC AND RANDOM ERRORS

Systematic Errors

Systematic errors produce a consistent difference from the true value. They can be corrected if the cause and value of the error are known. For example, times obtained from the watch that is always 5 minutes wrong can be corrected to give the true time.

Random Errors

Random errors give a lack of precision to any measurements we make. They may be minimised with careful attention to technique but will always be present to some degree. Random errors cannot be corrected but can be reduced by making multiple measurements and taking the average value.

Random error reduces by $\sqrt{n}$ of the number of measurements n made. So, for example, the average of 4 measurements will halve the random error.

Random errors will result from:

- Technique used to make the measurement.
- Intrinsic limits of ultrasound scanning, e.g. image resolution, geometric artefacts.
- Signal noise.
- Target movement.
- Introducing variables such as comparing measurements across machines.

ULTRASOUND MEASUREMENTS IN PRACTICE

We consider making measurements from images first then look at Doppler measurements.

The accuracy of measurements made from ultrasound images depends on the physical limitations of the ultrasound scanner and the process of image creation, and on the sonographer making the measurements. These two may be considered separately but the overall result will depend on the combined errors and limitations of both.

PHYSICAL CONSTRAINTS

The accuracy of measurement will depend on:

- Accurate scaling of the image
- Caliper accuracy
- Digital quantisation
- Image resolution

Scaling

In the lateral direction the scale relates to the physical width of the probe. In the case of a linear array, with no beam steering at the edge of the field of view, this is simply seen as the width of the active transducer array (Figure 10.2).

In the axial direction the scale depends on the **system velocity**. The positioning of an echo in the axial direction depends on the go-and-return time of the pulse which is determined by the speed of sound in tissue. For soft tissue the scanner assumes a system velocity of $1540\,\text{ms}^{-1}$. In reality the speed of sound varies by $\pm5\%$ around this value. This variation will lead to misplacement of echoes and limits the accuracy of any axial measurement. In some clinical situations a more precise system velocity is used that better matches the tissue being measured. For example, in ophthalmic scanning the speed of sound for the cornea and lens ($1641\,\text{ms}^{-1}$) and aqueous and vitreous humour ($1532\,\text{ms}^{-1}$) may be used instead of $1540\,\text{ms}^{-1}$. More recent developments that enable **speed of sound correction** improve the accuracy of measurement (see Chapter 13).

Caliper Accuracy

Caliper accuracy is the accuracy of the indicated caliper measurement compared with a known measurement in a $1540\,\text{ms}^{-1}$ tissue mimicking phantom. It is generally <1–2% (1 mm in 10 cm). Endoscopic probes tend to be slightly less accurate.

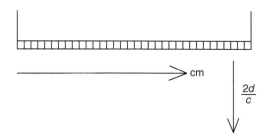

FIGURE 10.2 The dimensions of an image produced by a linear array.

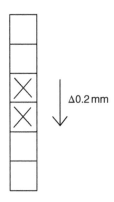

FIGURE 10.3   The quantisation error of a digitised caliper measurement.

Caliper precision is limited by **quantisation error** (Figure 10.3). As a caliper is moved from one pixel to the next there will be a discrete change in the value of the reading given. For example, 4.6–4.8 mm with $\Delta = 0.2$ mm. The exact value of this will depend on the pixel density, image depth and zoom used. It is therefore not meaningful to quote point to point linear measurements to a greater precision than this level of precision.

Image Resolution

The lateral and axial resolution of the image will limit the detail seen in the image and therefore limit the precision of measurements that may be made. Typical values for resolution are:

Axial resolution   ~1.5 mm at 3 Mhz
~0.6 mm at 7 MHz

Lateral resolution depends on beamwidth – typically 1.5–3 mm

The definition of edges in the image will also be affected by artefacts such as:

- Reverberation between closely separated boundary layers.
- Poor visualisation due to curved surfaces running in an axial direction.
- Clutter from slice thickness artefact, grating lobes and multiple reflections.
- Poor visualisation and image distortion due to patient habitus and overlying fatty tissue causing refraction and beam defocusing.

These factors are all intrinsic to the image creation and display process within the scanner.

SONOGRAPHER-BASED CONSTRAINTS

When making measurements the accuracy and precision of measurement will also be affected by sonographer choices when operating the scanner and placing the calipers.

Primary choices:

- **Probe frequency** – higher frequency gives finer resolution.
- **Focus** – set the focus to the level of the target of interest for best resolution.
- **Compound imaging** – improves the definition of curved surfaces.
- **Zoom** – write zoom before freezing the image gives more information, zooming the frozen image simply enlarges the information already present.

PRINCIPLES FOR MAKING RELIABLE MEASUREMENTS

Target Visualisation

Ensure the target is as clearly visualised as possible. Boundaries appear sharpest when viewed perpendicular to the ultrasound beam axis (Figure 2.3). Caliper placement will always be more reliable if placed on an interface imaged perpendicular to the beam and the ideal angle of approach to the target should be chosen so that this is the case. **Compound imaging** will improve the visualisation of curved surfaces.

Frame Freeze

The ultrasound probe must be held completely still when freezing the image to avoid blurr in the image (see Figure 7.27). This is especially the case when a lot of **frame averaging (persistence)** or **multiple transmit focal zones** (reducing frame rate) are being used. It may necessitate asking the patient to breath hold whilst freezing the image.

Use First Interface

It is generally a good principle to use the 'first interface' for caliper placement in making measurements (Figure 10.4). Immediately distal to the first interface the image may be confused with **reverberations** arising from first interface layers and **slice thickness artefacts**.

Measurements Protocols

In order to obtain reliable and robust measurements for clinical decision making it is good practice to use measurement protocols to ensure everyone making the

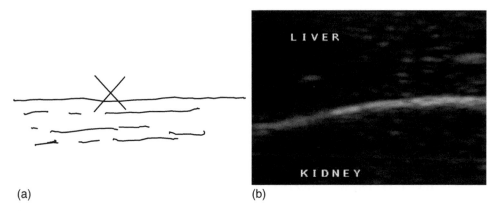

(a)   (b)

FIGURE 10.4   (a) Diagram of caliper placed on first interface of the target to be measured.
(b) Enlargement of liver-kidney boundary on an ultrasound image.

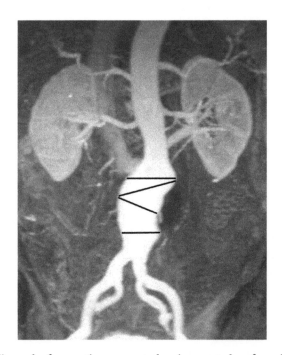

FIGURE 10.5   Radiograph of an aortic aneurysm showing examples of possible diameter
measurements.

measurement is making it repeatably in the same way. This is particularly important
where fine or difficult measurements are being made. Examples of protocol driven
measurement where image plane and caliper placement are specified include foetal
biometry, echocardiographic measurements and aortic aneurysm screening. [1–3]

As an example of the problem of deciding what to measure, the radiograph in
Figure 10.5 illustrates a number of ways in which calipers could be placed to gauge

the size of an aneurysm. A protocol should specify which measurement to use for consistency of reporting.

Actual Measurements

As shown in Figure 10.6, the measurement of distance between the two boundaries may vary between AF and CD depending on where the you place the calipers. Different sonographers are likely to place the calipers in slightly different positions, even given the principle of first interface placement. This is especially the case when the boundary is less distinct. This variation will give rise to a random error in the measurement.

In addition to the random variation in making such measurements, you will have a **personal systematic error** in the measurements you make. That is, you will generally tend to place the calipers in a particular spot within the image of a complex boundary. That will be slightly different to someone else, in other words, it is a personal preference particular to you.

Factoring Variability

These scanner-based factors and personal factors all contribute to the random error associated with any particular measurement. They give rise to intra- and inter-operator variability.

Random errors reduce as $\sqrt{n}$ of the number of measurements made, i.e. averaging 4 **independent measurements** of the same thing reduces the error associated with the measurement by half. Independent measurement in this case would be unfreezing the image, acquiring it again and re-making the measurement.

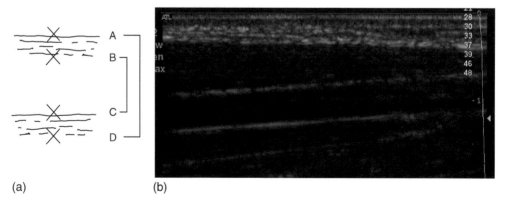

(a)   (b)

FIGURE 10.6   Diagram showing the possible range of caliper measurements across two interfaces (a). Illustration of such a pair of interfaces in a blood vessel (b).

For an area or volume measurement V, the random errors v, combine as shown:

$$\frac{v}{V} = \sqrt{\left(\frac{a}{A}\right)^2 + \left(\frac{b}{B}\right)^2 + \left(\frac{c}{C}\right)^2}$$

where a/A, b/B, c/C are the error and dimensions contributing to the volume V.

Ultrafine Measurements

Measurements of fine structures at the limit of resolution in the B-mode image may be successfully made where averaging of the measurement can be used. This may be by making multiple measurements and averaging. Or it may be by using some other averaging technique. An example of this is in the measurement of intima-media thickness (IMT) and flow mediated vessel dilatation, where the changes observed are of the same order as the axial resolution of the scanner.

For reliability and repeatability in all such ultrafine measurement, a **scan protocol** must be followed that includes a comprehensive set of machine settings to be used such as gain, zoom, frame averaging, position of focus, greyscale mapping etc.

In the case of IMT a longitudinal image of an artery is obtained such that the vessel wall with an intima-media boundary is visualised across the image (Figure 10.7). A dedicated linear cursor is placed along the intima-blood boundary and another along the intima-media boundary. Even though this boundary may not be sharply distinct along its length, the line of it can be seen and the linear cursor effectively averages the line of the boundary across the image. Some scanners provide an automated detection of the boundaries for IMT measurement.

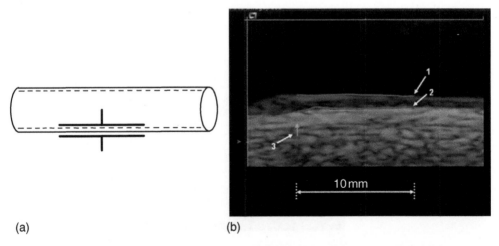

(a)   (b)

FIGURE 10.7   Diagram showing positioning of cursors for IMT measurement (a) and (b) an image with automated detection of the boundaries. Note the scale. The measurement is at the limit of ultrasound resolution. *Source:* Kastelein and de Groot [4]/with permission of Oxford University Press.

Ultrafine measurements of changes in vessel diameter with flow mediated dilatation or the cardiac cycle may also be made automatically using speckle tracking techniques. The RF echoes from the proximal and distal vessel wall-blood boundaries are tracked using autocorrelation to determine movement (see Chapter 9).

NOTES

- An ECG signal may be used to gate the measurement to the cardiac cycle.
- The measurement should be repeated several times and results averaged to improve precision.

MEASUREMENT OF CIRCUMFERENCE, AREA, AND VOLUME

Measurements involving more than one linear dimension are frequently required in ultrasound examinations. A number of approaches may be used in making such measurements. Each involve assumptions that result in the measurement being more or less accurate.

Direct Measurement

Area and circumference may be measured by directly tracing out the boundary to be measured. Using a tracker ball or other interface, the caliper is traced around the boundary, and the delineated circumference or area enclosed is calculated.

- Relies on hand-eye coordination of the sonographer tracing round the area.
- For circumference errors of overshooting or undershooting the boundary add to the length increasing overall error.
- For area measurement, overshooting and undershooting tend to even out.

Assume a Simple Shape

It is often sufficient to assume a simple shape such as an ellipse and match that shape visually to the target to be measured.

- Many targets approximate an ellipse very well.
- Reduces input required by the sonographer.
- Good reproducibility.

In the case of an elliptical shaped target the estimate may be made by measuring the long and short axis of the ellipse so the scanner can calculate the values, e.g. for foetal head measurements (Figure 10.8). The placement of an elliptical cursor on the target will perform this calculation.

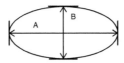

FIGURE 10.8   The dimensions used when assuming an ellipse for calculating area or circumference.

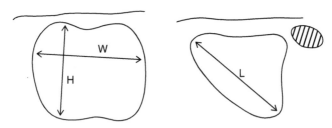

FIGURE 10.9   The measurements to be made for calculating bladder volume.

Use a Simplified Model of the Target

By using a simplified model of a complex shape, an estimation of the circumference, area or volume may be made using just a few linear measurements.

An example of such a model may be used to measure bladder volume (Figure 10.9) [5].

One model used is that of a prolate ellipsoid. The formula used is

$$V = 0.52 \times \left(\text{length} \times \text{width} \times \text{height} \right)$$

and volume $= V \pm 25\%$.

This degree of imprecision may be considered adequate for this measurement.

Slicing Models

A closer estimate of the true volume may be made by measuring a series of slices through a volume and calculating the volume as a series of trapezoid volumes added together as shown in Figures 10.10 and 10.11.

Surface Segmentation and Automated Methods

With the advent of 3D ultrasound and artificial intelligence algorithms, it is possible to automatically detect surfaces within the image data and **segment** or define the volume to be measured (see Chapter 12).

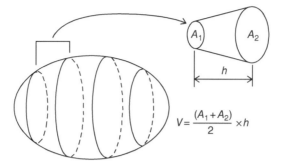

FIGURE 10.10   The method of calculating volume using trapezoid slices.

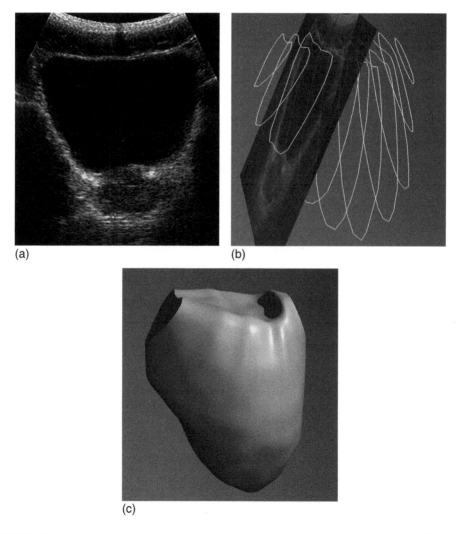

(a)

(b)

(c)

FIGURE 10.11   Illustrating the principle of deriving the bladder volume from a series of image slices. (a) A B-Mode image slice, (b) illustrates the 3D position of the slices to be used and (c) shows the reconstructed bladder volume. *Source:* Images courtesy of Dr Graham Treece.

DOPPLER WAVEFORM MEASUREMENTS

Doppler Set-Up

When observing Doppler waveforms and in order to make reliable velocity measurements that maximise the available information in the sonogram, it is important to set up the Doppler display correctly.

Image/Sonogram Size

When viewing Doppler waveforms, the sonogram is the main focus of interest and the sonogram display should be the major part of the screen display with the B-mode image reduced in size to be used for guidance only.

Scale

- Set the velocity scale so that the waveform fills the space. You then have the largest size of display from which subtle changes in waveform may be seen and more precise measurements made (Figure 10.12).
- Set the time-base 4–6 seconds or to see what you want to see and no more. For example, to measure systolic rise-time (SRT) a faster time base is needed (2–3 seconds) and to look at autonomic changes in diastolic flow a longer time base is needed (e.g. 15 seconds) (Figure 10.13)

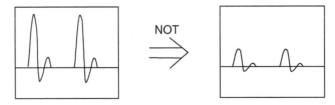

FIGURE 10.12   The correct scaling to use when viewing a Doppler waveform.

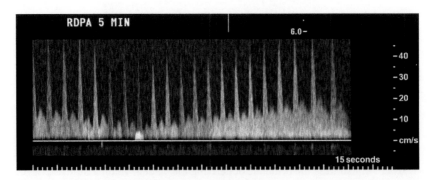

FIGURE 10.13   Using a 15 second time base to illustrate the cycling of sympathetic tone in an artery.

Gain

Doppler gain should be adjusted so that the greatest intensity seen in the waveform *just* reaches peak white (see Figure 8.26)

NOTE

Over-gain, overestimates velocity
Under-gain, underestimates velocities

Freeze the Image

Whilst acquiring Doppler waveforms, the B-mode image should be frozen to avoid time sharing Doppler with B-mode. This gives much better-quality waveforms as it uses longer, more frequent samples, giving better frequency and time resolution on the sonogram.

Weak Signals

To improve imaging where the signal is weak:

- Reduce beam steer toward 'straight ahead' and **'heel and toe'** the probe to obtain a good Doppler angle (recall Chapter 3 – a forward looking beam is more sensitive) See Figure 9.9.
- At very shallow angles there is poorer penetration of ultrasound into the vessel lumen and errors in velocity measurement can occur. Therefore, use an angle of >45° (Figure 10.14)

Doppler Angle

The greatest source of error in making measurements from Doppler waveforms is a misalignment of the **Doppler angle cursor**. At an angle of 60° a 2° error in cursor alignment will produce an error in velocity measurement of ~6% (see Figure 8.25). The angle cursor moves in discrete steps, e.g. 2° steps, giving a **quantisation error** (Figure 10.15). Accurate alignment of the angle cursor is therefore vital for reliable velocity measurements.

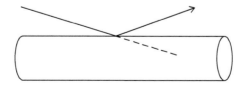

FIGURE 10.14 Reflection of the Doppler beam at the vessel wall with a shallow angle of insonation giving poor penetration into the vessel.

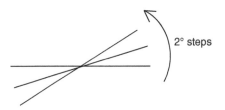

2° steps

FIGURE 10.15   Typical quantisation error in positioning the angle cursor.

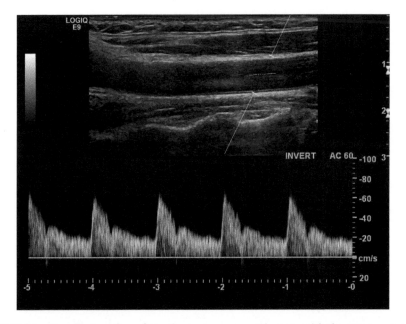

FIGURE 10.16   Doppler waveform from the common carotid artery with the artery correctly aligned along its length to show the vessel walls and intima and the Doppler angle set correctly.

For **cardiac work**

- It is usually possible to obtain an alignment for the Doppler signal that is in line with the flow being measured, e.g. through a valve. In this case the Doppler angle is 0°.
- When looking at flow jets the Doppler beam must be accurately aligned with the jet flow for a 0° angle assumption to be valid (see Figure 12.15).

For **vascular work**, in order to make accurate measurements in arteries and veins the following should be observed:

- Align the vessel so as to show the longest length of sharply imaged vessel walls across the width of the B-mode image, as shown in Figure 10.16. Then you know you are lined up with the vessel axis and in line with the flow in most cases.
- When the vessel is very tortuous then use the tangent to the curve of the vessel at the point of measurement to position the angle cursor (A) (Figure 10.17).

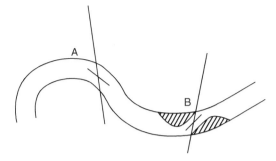

FIGURE 10.17   Diagram showing the correct alignment of the angle cursor in the case of a curved vessel (A) and an eccentric jet (B).

- Where disease causes an eccentric jet of flow then align the angle cursor to the line of the jet (B).

An ideal Doppler angle is 45–60° – (or fix at 60° to minimise variability due to angle effects)

The **Doppler sample volume** should be wide enough to ensure that the peak velocity in the flow is captured. To ensure this and reduce **intrinsic spectral broadening**, a sample volume that straddles the vessel is recommended.

Having obtained the correctly aligned best waveform possible, measurements may be made using the **manual cursor** or **auto trace** as discussed in Chapter 8.

WAVEFORM INDICES

These were used extensively in the early days of CW scanning to avoid the 'unknown angle' problem (i.e. no B-mode image, just the Doppler sonogram). They are still used in some situations. [6]

Pulsatility Index (PI)

Pulsatility Index (PI) – gives a measure of how pulsatile a waveform is.

$$PI = \frac{(\text{peak} - \text{peak})}{\text{average of peak envelope}}$$

> **NOTE**
> If the equation is written out in terms of the Doppler equation to calculate the velocities, then $\cos\theta$ appears on top and bottom so cancels out, so PI is independent of Doppler angle.

As the waveform becomes less pulsatile the PI reduces in value. (PI value range ≥ 0)
Examples as shown in Figure 10.18:

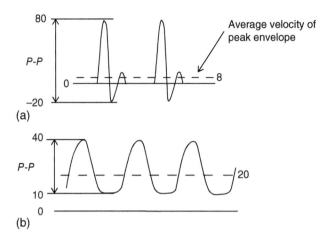

FIGURE 10.18   Examples of pulsatility index parameters. (a) 80/8 = 12.5, in a normal femoral waveform and (b) 40/20 = 1.5, in a damped waveform.

Resistance Index (RI)

Resistance index (RI) – also known as **Pourcelot's RI** – gives a measure of peripheral resistance

$$RI = \frac{(A - B)}{A}$$

Diastolic flow is a good measure of peripheral resistance and as resistance reduces B increases relative to A and the value of RI decreases. (RI value range 0–1)

NOTE

Again, $\cos \theta$ appears on top and bottom so cancels out, so RI is independent of Doppler angle.

Examples as shown in Figure 10.19:

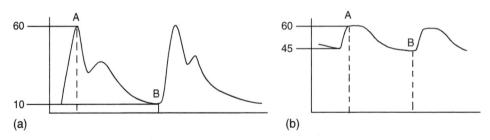

FIGURE 10.19   Examples of Resistance index parameters. (a) (60 – 10)/60 = 0.83, in a normal common carotid waveform and (b) (60 – 45)/60 = 0.25, in a low resistance fistula waveform.

A/B Ratio

A/B ratio – is similar to RI with the same A and B values uses as shown above. This index is still used in obstetrics

COLOUR DOPPLER ULTRASOUND

CDU is used to show where there is flow and can give a good indication of where to place the Doppler sample volume to obtain the Doppler waveforms. However, it is of limited use in making measurements in its own right. Velocity measurements cannot be made from CDU as the flow angle is not known.

CDU should only be used to make measurement of patent lumen diameter when there is an inadequate view on the B-mode image. This is because CDU suffers from bleed into the vessel wall or dropout adjacent to the vessel wall and so is not as accurate as measuring from a sharply imaged wall on B-mode.

As an example of combining measurements involving B-mode and Doppler waveforms we look at the measurement of volume flow.

MEASUREMENT OF VOLUME FLOW Q

Volume flow is potentially a very useful measurement to make as it is a measure of the oxygen-carrying capability of a vessel (Figure 10.20). However, its use is limited due to the way errors accumulate in its measurement and the assumptions that must be made. [7]

Cross-Section Area A

For an artery, the assumption is that the cross section is circular. The area A may therefore be calculated by measuring the diameter of the vessel lumen from the B-mode image. A longitudinal image is obtained clearly showing the vessel wall with the intimal layer seen. The lumen diameter D is then measured. Using a

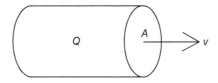

FIGURE 10.20   Definition of volume flow as the volume of a cylinder Q given by the cross-sectional area A times the average velocity of blood v.

Volume flow $Q = v. A$ where v is the average velocity of blood,
                A is the cross-sectional area of the vessel

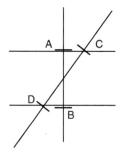

FIGURE 10.21   Correct cursor position AB and inaccurate positioning CD for diameter measurement of vessel.

cross-sectional view to measure the diameter may give an error if the cross-section is not truly perpendicular to the line of the vessel, such that CD was being measured (Figure 10.21).

$$A = \pi \frac{D^2}{4}$$

Any error in D will be squared when calculating the area.

Average Velocity

With the Doppler angle cursor accurately aligned, a waveform is obtained and the average velocity is estimated from the average velocity function on the scanner. In order to obtain the most reliable estimate the following must be observed:

- For a Doppler waveform that has reverse flow or reduces to zero for part of the cardiac cycle, ensure that the average velocity curve accurately maps the flow as reverse or zero at these points (see average velocity in Chapter 8).
- Ensure the sample volume straddles the vessel so all velocities across the lumen are included as shown on Figure 10.22. However, this is a limited measurement

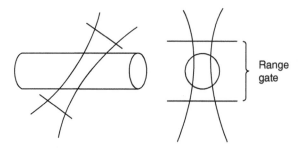

FIGURE 10.22   The correct setting of the range gate (a) and (b) showing the problem of the focus on the Doppler beam catching the centre line velocities and missing the low velocities at the lateral vessel walls.

because the Doppler ultrasound beam is focused at the sample volume depth. Even with the sample volume straddling the vessel, the higher velocities in the centre of the vessel will be emphasised over the slower velocities near the vessel walls. Depending on the velocity profile across the lumen, this may lead to up to 30% over estimate of the average velocity.

- The measurement must be made over a complete number of cardiac cycles to ensure flow throughout the cardiac cycle is equally represented, for example, end diastolic to end diastolic over 3–4 cycles.
- The waveform is then obtained and the average velocity measured over a complete number of cardiac cycles, for example, 3 cycles end diastole to end diastole as shown in Figure 10.23.
- Using the fact that measurement error reduces as $\sqrt{n}$, the estimate of volume flow may be improved by repeating the measurements and averaging the result.

To give an idea of the precision of volume flow measurement, in a brachial artery diameter 6 mm, the volume flow Q may be estimated as $Q \pm 12\%$. This is good enough when considering the very high flows seen in arteriovenous fistula used for haemodialysis.

Better estimates of volume flow are possible using ultrafast ultrasound techniques (see Chapter 13).

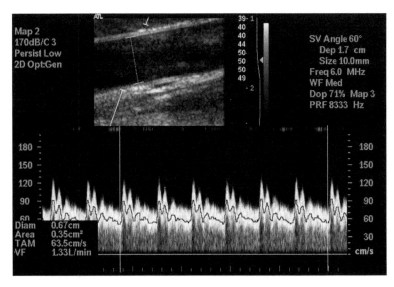

FIGURE 10.23   An example of volume flow measurement in a renal arteriovenous fistula showing diameter measurement perpendicular to vessel walls, and average velocity measured over a complete number of cardiac cycles.

REFERENCES

1. Public Health England (2021). https://www.gov.uk/guidance/fetal-anomaly-screening-programme-overview (accessed August 2022).

2. ISUOG (2019). Practice guidelines: ultrasound assessment of fetal biometry and growth. *Ultrasound in Obstetrics & Gynecology* 53: 715–723. https://doi.org/10.1002/uog.20272.

3. Benson, R.A., Meecham, L., Fisher, O. et al. (2018). Ultrasound screening for abdominal aortic aneurysm: current practice, challenges and controversies. *The British Journal of Radiology* 91 (1090): 20170306. https://doi.org/10.1259/bjr.20170306.

4. Kastelein, J.J. and de Groot, E. (2008). Ultrasound imaging techniques for the evaluation of cardiovascular therapies. *European Heart Journal* 29: 849–858. https://doi.org/10.1093/eurheartj/ehn070.

5. Dicuio, D., Pomara, G., Menchini Fabris, F.M. et al. (2005). Measurements of urinary bladder volume: comparison of five ultrasound calculation methods in volunteers. *Archivio Italiano di Urologia, Andrologia* 77: 60–62.

6. Oates, C. (2001). *Cardiovascular Haemodynamics and Doppler Waveforms Explained*. Cambridge: Cambridge University Press.

7. Oates, C.P., Williams, E.D., and McHugh, M.I. (1990). The use of a Diasonics DRF400 duplex ultrasound scanner to measure volume flow in arterio-venous fistulae in patients undergoing haemodialysis: an assessment of measurement uncertainties. *Ultrasound in Medicine and Biology* 16: 571–579.

CHAPTER 11

Safety and Quality Assurance

As with anything we do with patients, in performing diagnostic ultrasound scans, we must ensure that we operate in a safe manner that recognises all the potential hazards and minimises the risks. A **hazard** is anything that can cause harm and **risk** is the chance that any hazard will actually cause somebody harm. An additional factor to be considered in the use of a modality such as ultrasound is the benefit of performing the examination to the risk it incurs. Potential hazards may range from severe (e.g. injury to an unborn child) to very minor (e.g. clothing is soiled by the ultrasound gel used). In the case of therapeutic or surgical use of ultrasound, the specific aim is to cause a change to the tissues being insonated. For diagnostic usage, such changes must be avoided.

Likewise, the risk may be high or low, so a procedure that has a high risk of a severe hazard is not one we would routinely perform! The risk versus benefit is also an important consideration and, for example, comes into the debate about whether it is right to perform ultrasound solely for souvenir scans.

It is with these considerations in mind that we approach our discussion of the safety of diagnostic ultrasound.

The first thing to note is that in exposing the body to ultrasound we are putting energy into tissue. As we saw in the second chapter, apart from the small proportion of energy reflected back to the transducer as echoes, the rest is absorbed by the tissue as specular reflection and scattering take place. In fact, at the low megahertz frequencies used in diagnostic ultrasound the proportion of energy directly absorbed by tissue is much higher than the proportion scattered, which is less than 10%. That which is scattered is also eventually absorbed. The deposition of energy will, in general, lead to tissue heating. In addition, since the energy is carried in the form of

Ultrasound Technology for Clinical Practitioners, First Edition. Crispian Oates.
© 2023 John Wiley & Sons Ltd. Published 2023 by John Wiley & Sons Ltd.

mechanical vibrations of molecules as they experience the sound wave, we may antic-ipate that there will be mechanical effects within the tissue that may be hazardous. Throughout this section reference has been made to *The Safe Use of Ultrasound in Medical Diagnosis* [1].

ENERGY, POWER, AND INTENSITY

As we saw in Chapter 1, the **energy** in a sound wave is proportional to the acoustic pressure p squared.

$$E \propto p^2 \text{ measured in Joule} \left(J \right)$$

Power is the rate at which energy is transferred

$$W = \frac{E}{t} \text{ measured in Js}^{-1} \text{ or Watts} \left(W \right)$$

Intensity is the power per unit area, or the energy flowing across an imaginary surface cutting across the ultrasound beam. The higher the intensity, the greater the energy transferred to that point in tissue (Figure 11.1).

$$I = \frac{W}{\text{area}} \text{ for diagnostic ultrasound this is measured in mWcm}^{-2}$$

Intensity is also proportional to the acoustic pressure p squared. For a plane wave, the instantaneous intensity is given by

$$I = \frac{p^2}{Z} \text{ where } Z = \rho c, \text{ the acoustic impedance}$$

So, at regions of high intensity the pressure amplitude p of the sound wave will be greatest, and hence particle displacement and mechanical stress will be relatively large.

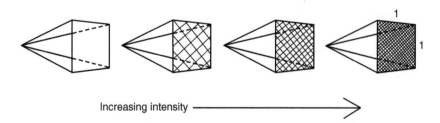

Increasing intensity ⟶

FIGURE 11.1   Diagram illustrating the effect of increasing intensity at a point in tissue.

Intensity is therefore the key measure in relation to consideration of any hazard caused by insonating tissue.

MEASURING INTENSITY

Acoustic pressure can be measured with a **hydrophone**. This is a very small calibrated ultrasound transducer of known area (e.g. $0.5\,\text{mm}^2$) that can pick up the signal from one point within an ultrasound beam. The amplitude of the signal from the hydrophone transducer is proportional to the acoustic pressure; hence, the intensity at that point can be calculated by squaring, or otherwise measuring, the power of the signal as shown in Figure 11.2.

INTENSITY

In general, diagnostic ultrasound uses **pulses** of ultrasound to interrogate the target. Different modalities (e.g. B-mode, CDU, and Doppler) and different machine settings within each modality will change the pulse shapes and the pulse repetition frequency. Each pulse delivers energy to the tissue and in between pulses the amplitude is zero. We therefore need to carefully define what intensity we are measuring.

Within a complex beam shape produced by a transducer, the intensity of the sound wave varies from point to point, as seen on an intensity contour map (Figure 11.3). At a given field point within the field of view, insonation is intermittent as it comes in pulses – both in terms of individual transmitted pulses and 'pulses' at the field point P as the beam sweeps past the field point (Figure 11.4).

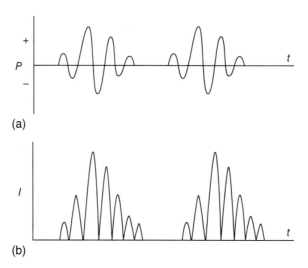

FIGURE 11.2 As acoustic pressure amplitude (a) oscillates between positive and negative values, amplitude squared, the intensity, has a positive value (b) throughout the cycle.

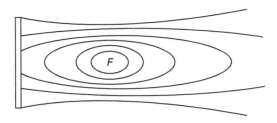

FIGURE 11.3   Schematic intensity contour map of a focused beam. Highest intensity is at the focus *F*.

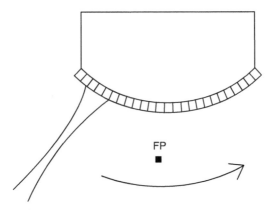

FIGURE 11.4   A field point *P* experiences a 'pulse' of ultrasound as the beam is scanned across the field of view as the individual pulses are fired in that direction.

In order to assess where the greatest risk of harmful effects might be, we consider how the intensity varies in 'space', i.e. within the volume of tissue insonated, and how it varies in 'time' as ultrasound beams move across the field of view and are only intermittently experienced at a particular field point. A number of defined measurements are useful (see Figures 11.5 and 11.6).

The **spatial peak intensity** is the highest intensity found in the transmitted ultrasound beam. It is usually at the focus. The **spatial average intensity** is the average value across the width of the beam.

The **temporal peak intensity** is the highest intensity found within a single pulse, the pulse peak intensity. The **temporal average** intensity is the intensity averaged across the whole pulse cycle including the dead time between pulses. The **pulse average intensity** is the intensity averaged across the pulse length, not including dead time.

Using these definitions, the following values are defined.

I_{SPTA} **spatial peak temporal average intensity**. This is the highest value of intensity in the ultrasound beam averaged over many pulses. In the case of a swept beam this is averaged over many fields. This measure is the best indicator of potential heating of tissue by the ultrasound.

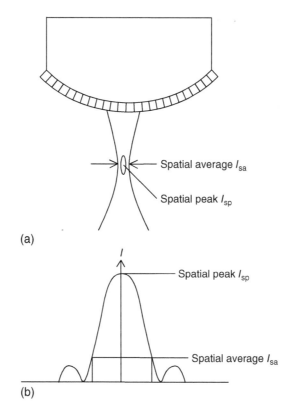

(a)

(b)

FIGURE 11.5 Definition of spatial peak and spatial average intensity for an ultrasound beam (a) in plan view of an ultrasound beam and (b) in cross-section.

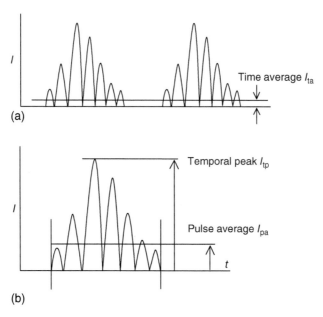

(a)

(b)

FIGURE 11.6 Definition of (a) temporal peak intensity and (b) temporal average and pulse average intensity.

I_{SPTP} **spatial peak temporal peak intensity**. This is the highest intensity in the ultrasound beam at the time when it is at maximum, i.e. at the time of peak pulse amplitude. It is the greatest instantaneous intensity in the field of view.

I_{SPPA} **spatial peak pulse average intensity**. This is the intensity averaged over the length of a pulse, not including dead time, at the point in the field of view where intensity is highest.

I_{SATA} **spatial average temporal average intensity**. This is the intensity averaged over the whole beam cross-section and over time at a particular distance from the transducer. It is the average intensity experienced at a particular field point.

NOTE

The spatial peak intensity will typically be at a transmit focus.

FACTORS AFFECTING DAMAGE POTENTIAL

The potential for ultrasound to damage tissue depends on a number of factors. These can be summarised as follows:

- **Transmit power** – increasing transmit power increases the energy input to the tissue.
- **Scan mode** (B-mode, Doppler, etc.) – affects pulse shape and pulse length, pulsing regime used, pulse repetition frequency, and distribution of pulsing within the field of view. In B-mode the beam is swept across the field of view and at any one point there will be a significant amount of time when there is no incident ultrasound, so the average intensity will be low. For Doppler modes and M-mode, ultrasound is repeatedly transmitted along the same line so the average intensity will be higher.
- **Frequency** – attenuation and mechanical effects are frequency dependent.
- **Focus** – the position and strength of each transmit focus.
- **Tissue type** – attenuation varies between tissues and sensitivity to injury for some tissues is of greater significance than others. Particularly sensitive, for example, are the lens of the eye and early pregnancy when the whole embryo can be within the sound beam and whole-body development is taking place.
- **Dwell times** – how long a single target point is within the ultrasound field. This will be greater for non-scanning modes such as Doppler or M-mode, where the ultrasound beam is repeatedly fired along the same line, but also depends on how long the probe is held in one position insonating a particular target.
- **Additional factors** – for example, the presence of bubbles when a contrast agent is used.

We will consider these factors in relation to thermal, and mechanical or non-thermal effects (Figure 11.7).

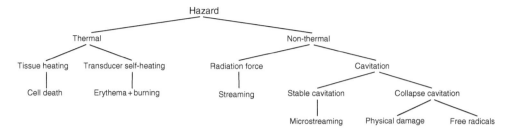

FIGURE 11.7 The relationship between hazards of ultrasound exposure showing physical mechanisms and mechanisms of physical and biological damage.

THERMAL EFFECTS

Normal human core temperature is accepted to be 37°C with a diurnal variation of ± 0.5–1°C.

The potential for damage increases with temperature rise and is related to the length of time an elevated temperature is sustained. In the case of foetal tissue, increases of $T > 2$°C have been shown to produce teratogenic effects in animal models. Certain periods of foetal development have been shown to be more sensitive to thermal insult than others, and in general, sensitivity to thermal insult is greater when tissues contain a large component of actively dividing cells.

The degree of heating within a given tissue will depend on a number of factors. The main factor is the **absorption coefficient** of the tissue (see Chapter 2). Generally, more dense tissues such as bone have a greater absorption coefficient and will therefore be heated to a greater extent (Table 11.1). In the case of bone, this may be as much as 50 times greater than most soft tissue. For soft tissue, absorption is the major component of attenuation of ultrasound accounting for some 90% with scattering losses accounting for no more than 10%.

TABLE 11.1 Attenuation coefficients of different tissues

Tissue	Attenuation coefficient α	
Blood	0.15	
Fat	0.6	
Brain	0.8	
Liver	0.6	
Kidney	1.0	
Muscle (along fibres)	1.3	Anisotropy
(Across fibres)	0.7	
Water	0.02	
Cortical bone	20	

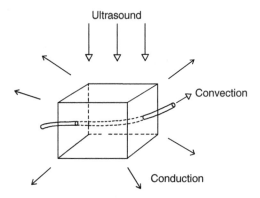

Ultrasound

Convection

Conduction

FIGURE 11.8   Mechanisms of energy transport in insonated tissue.

As shown in Figure 11.8, the degree of heating will depend on the balance between the energy input by ultrasound and the dissipation of that heat into surrounding tissue. Dissipation will be by conduction of heat into adjacent tissue and convection where there is perfusion by blood vessels that can carry the heat away. In terms of intensity measurement, the key measure for heating is the I_{SPTA}.

In the case of bone, the high coefficient of absorption can lead to significant localised increases in temperature at the bone surface that can raise the temperature of adjacent soft tissue to levels considered unsafe.

In order to give the user an indication of possible temperature rise from a given machine setup, the **thermal index** has been developed and is displayed on the image display screen.

THERMAL INDEX (TI)

Thermal index is defined as the ratio of the current acoustic output power from the transducer (W_0) to the power required to raise the tissue by 1 °C (W_{deg}).

$$TI = \frac{W_0}{W_{deg}}$$

So, a TI of 1 is indicating the possibility that the current output may raise the temperature within tissue by 1 °C.

This is based on a model with a very low level of attenuation, $0.3\,dBcm^{-1}MHz^{-1}$ (soft tissue $= 0.53\,dBcm^{-1}MHz^{-1}$). In other words, the beam is assumed to be more intense at depth than it actually is, to give a built-in safety margin.

Because of the possibility of enhanced heating when bone is in the field of view, three settings of the thermal index have been developed corresponding to different combinations of soft tissue and bone commonly encountered during scanning.

TIS

TIS – for soft tissue.

TIB

TIB – for imaging in a non-scanning mode (i.e. with a static beam, for example, pulse Doppler or M-mode) when bone is at or near the focus.

TIC

TIC – for when bone is near the skin surface, e.g. postnatal cranial scanning.

NOTE

It is important that the correct version of TI is selected for display for the tissue configuration you are scanning.

NOTES

- TI does not include a dwell time element and for obstetric scanning, guidelines shown in Figure 11.9 should be adhered to.
- TI indicates a relative risk rather than an absolute risk. That is, a scan performed using a TI of 2 compared to that using a TI of 1 represents a greater risk to the patient, although the absolute risk is not known.

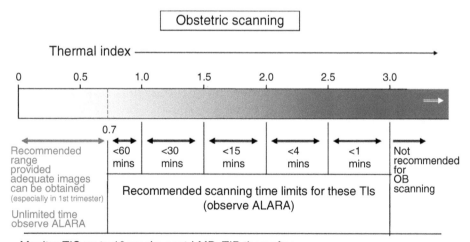

FIGURE 11.9 BMUS recommended maximum dwell times for increasing TI for obstetric scanning. *Source:* Haar [1].

- The TI will change in response to changing the scanner controls.
- TI will tend to be higher in pulsed Doppler mode, where pulses are longer than for B-mode and the beam is directed continuously at the same target. This is especially the case for TIB and TIC where bone is included in the field of view.
- There is likely to be an increased risk to the foetus of any temperature increase due to insonation by ultrasound when the maternal core temperature is already elevated, for example, due to maternal fever.

IMPORTANT

Note where on the display the current value of TI is shown and ensure you are operating within safe limits for the target of interest.

TRANSDUCER SELF-HEATING

When a transducer transmits acoustic energy, the transfer of electrical energy into transmitted ultrasound is not 100% efficient and the transducer will itself heat up. This is known as **transducer self-heating.** It arises from both electrical resistive heating in the transducer circuitry and from absorption within the transducer of the acoustic energy generated. Matching between the transducer and the external medium is not perfect and there remains a significant impedance mismatch between the transducer and the external medium (see Chapter 4). In the case of a transducer running in air, i.e. not in contact with the body, almost 100% of the ultrasound generated is reflected back into the transducer at the air interface. If the temperature of the front surface of the transducer becomes elevated, there is a risk to the patient when the probe is placed on the skin.

It is often transducer self-heating that restricts the acoustic power output of a transducer rather than a limit on I_{SPTA} or MI (see below).

The manufacturers safety standard, IEC60601-2-37(2007) [2] limits the transducer surface to <50°C when running in air and to <43°C when in contact with the patient.

NOTE

It is important to be aware and ensure that the probe surface, whether an external probe or an endo-probe, is not getting too hot (43°C would be a hot shower, 50°C will feel burning and too hot for a handwash).

NONTHERMAL EFFECTS

The causes of nonthermal effects of ultrasound in tissue have a number of mechanisms that may be divided into two groups - **non-cavitational** and **cavitational** mechanisms.

RADIATION FORCE

As a result of energy being absorbed by tissue, the tissue experiences a force called **radiation force** in the direction of propagation of the ultrasound. The time average value of this force per unit volume of tissue F_V is given by

$$F_V = \frac{2\alpha I}{c}$$

where α is the absorption coefficient, I the intensity of ultrasound and c the speed of sound.

The total force F_T exerted on the tissue depends on the total power W absorbed from the ultrasound beam.

$$F_T = \frac{W}{c} \quad \text{where } c \text{ is speed of sound}$$

Measurement of this force in water may be used to determine the acoustic power output of a transducer. For example, using a **radiation force balance** with a total acoustic absorber, the radiation force from incident ultrasound is balanced against a weight under gravity to determine the acoustic power (Figure 11.10).

STREAMING

When ultrasound passes through a liquid, the molecules and any suspended particles are free to move and radiation force will produce the phenomena of **streaming**. This

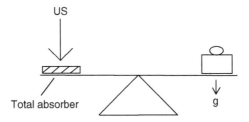

FIGURE 11.10 Diagrammatic illustration of a radiation force balance.

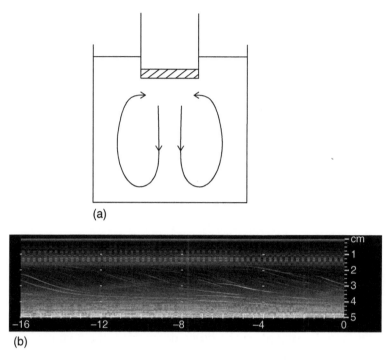

(a)

(b)

FIGURE 11.11   (a) Illustration of the movement within a fluid generated by radiation force from an ultrasound beam. (b) Shows an M-mode scan of bubbles in a beaker full of water. The bubbles are pushed down by the radiation force from the linear array transducer.

may sometimes be seen within cysts and other fluid-filled cavities in the body, showing up as motion on a CDU display (Figure 11.11).

In normal scanning, radiation force does not present a significant hazard but it is the force used to probe tissue in some modes of elastography (see Chapter 14).

CAVITATION

An acoustic sound wave is a pressure wave that alternately compresses and stretches the molecular structure through which it passes. Where there are small (microscopic) gas bodies or bubbles within the medium, they will oscillate in the sound field. At low intensities they will continue to oscillate and the phenomena is known as **stable cavitation** or **non-inertial cavitation**. Around the oscillating bubble micro-streaming can occur subjecting cells and sub-cellular structures to high and potentially damaging velocity gradients as the fluid slows down away from the bubble (Figure 11.12).

As the amplitude of the sound wave increases (higher intensities) the bubbles can grow by a process called **rectified diffusion** and then catastrophically collapse within the space of one cycle of the ultrasound wave due to the surrounding tissue

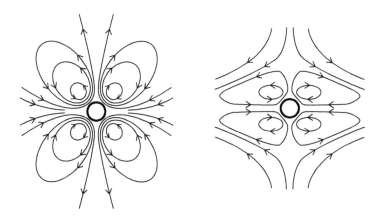

FIGURE 11.12 Two patterns of microstreaming that may occur around microbubbles. *Source:* Adapted from Jalal and Leong [3].

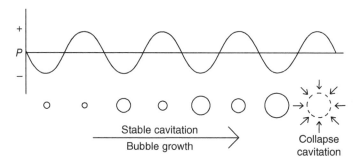

FIGURE 11.13 Illustration of bubble growth followed by collapse in an oscillating sound field.

bearing down on it as the pressure increases as shown in Figure 11.13. This is called **inertial** or **collapse cavitation** and is the main non-thermal damage mechanism in tissue.

Rectified Diffusion

Microbubbles are able to grow in an oscillating sound field by a process known as **rectified diffusion** (Figure 11.14). The bubble will shrink and expand as the acoustic pressure changes. When the pressure is low – a phase known as **rarefaction** – the bubble expands and has a larger surface area through which gas dissolved in the surrounding tissue can diffuse. When pressure grows during the **compression** phase of the cycle gas in the bubble is squeezed and will diffuse out of the bubble back into the surrounding tissue. However, in that phase the bubble has a smaller surface area, so the amount of gas leaving during compression is less than the gas entering during rarefaction. The net result is that with each compression-rarefaction cycle the bubble grows.

(continued)

(continued)

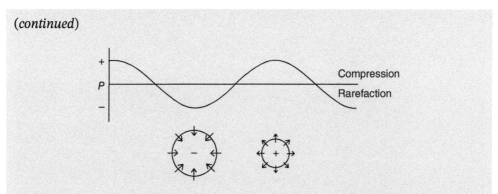

FIGURE 11.14   The movement of gas vapour into and out of a microbubble oscillating in the sound field.

As a bubble collapses the pressure and temperature within the bubble increases dramatically and can briefly reach extremely high values (up to 2000 atm and 5000 °C) in a highly localised space. This causes large mechanical stresses on the surrounding medium and potentially creates **free radicals** that can cause chemical damage. Free radicals are highly reactive molecules created by stripping off electrons from stable molecules. The potential damage to tissue includes disruption of cell membranes, haemorrhage of small vessels, and chemical damage to important molecules such as DNA.

Whether cavitation is possible or not depends on a number of factors. Firstly, it is more likely if there are existing free gas bodies within the tissue that can act as **cavitation nuclei** on which the cavitation mechanism can commence. Some tissue has more such cavitation nuclei than others, especially if it is adjacent to gas-filled spaces, for example the lung and the gut.

The pulse shape and frequency of the ultrasound will affect the likelihood of cavitation. Whereas thermal effects depend on the average energy input to the tissue, cavitation depends on the amplitude and length of individual pulses. In particular, it is the magnitude of the **negative pressure** experienced during the rarefaction phase of the sound wave cycle that determines the likelihood of collapse cavitation (Figure 11.15). It is the transition from a large negative pressure to a positive pressure that causes the catastrophic collapse of the bubble as the surrounding medium pushes in on it.

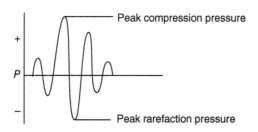

FIGURE 11.15   Definition of peak compression and rarefaction within an ultrasound pulse.

MECHANICAL INDEX (MI)

Corresponding to the thermal index shown on the ultrasound image display, a mechanical index has been developed to indicate the potential for collapse cavitation to occur in-vivo. It is based on measurements in water that have been de-rated assuming an attenuation coefficient of $0.3\,dBcm^{-1}MHz^{-1}$. As this is less than that of soft tissue ($0.53\,dBcm^{-1}MHz^{-1}$) it will overestimate the likelihood of cavitation occurring and so has a built-in safety margin.

$$MI = \frac{p_{r,3}}{\sqrt{f_c}}$$

where $p_{r,3}$ is the peak rarefaction (p_-) pressure in water assuming an attenuation coefficient of $0.3\,dBcm^{-1}MHz^{-1}$ and f_c is the centre frequency of the ultrasound pulses.

NOTES

- MI is given without specifying units, i.e. it is a numerical index.
- The FDA currently limits scanner output to a maximum MI of 1.9.
- Below an MI of about 0.4 bubble growth is unlikely even if gas bodies are present in the tissue.
- Lower frequencies will give higher values for MI.

IMPORTANT

Note where on the display the current value of MI is shown and ensure you are operating within safe limits for the target of interest.

ALARA

As Low As Reasonably Achievable

The ALARA principle is: to work so that any risks are as low as reasonably achievable.

- Use the minimum transmit power that will give the images you need.
- Reduce dwell time, especially to critical organs, to that necessary to obtain the information required.

CONTRAST AGENTS

Contrast agents (CA) used in **contrast enhanced ultrasound (CEUS)** consist of injectable liquids with suspended gas bodies (see Chapter 12). It is the suspended gas bodies that are highly reflective to ultrasound, enabling them to show up in low concentrations and when visualisation is otherwise poor. However, the use of such contrast agents introduces two distinct areas of risk for harm.

- The **pharmacological risk** associated with the use of an injectable substance.
- The **risk of physical effects** arising from the introduction of gas bodies into tissue that is insonated.

Clinical Risk

Contrast agents are not usually intended to be pharmacologically active but, as with any substance injected into the body, reaction such as anaphylaxis is a possibility. Contrast agents should always be used in accordance with the manufacturer's information on the agent and its administration, and knowledge of the patient.

Bio-effects from Ultrasound

With the administration of the contrast agent, gas bodies (bubbles) are introduced into the tissue being insonated. The threshold intensity for cavitation in soft tissue is generally very high but the presence of CA bubbles makes the potential risk of cavitation much greater.

The potential bio-effects of CA use depends on the following factors:

- **Rarefactional pressure amplitude** (P_-).
- **Agent dose** – multiple doses of CA will increase risk and the bio-effects seen.
- **Agent delivery** – for example, whether the CA is given as a bolus injection or more slowly.
- **Imaging mode** – affects pulse shape and amplitude
- **Tissue properties** – some tissues are more sensitive to cavitational damage than others (e.g. kidney)
- **Clearing pulses (flashing)** – For some investigations a 'clearing pulse' may be used that deliberately bursts the bubbles via collapse cavitation so as to look at, for example, refill times in perfusion studies. Such a clearing pulse has a high amplitude with a large rarefactional pressure amplitude that is deliberately designed to destroy the bubbles.

THEREFORE NOTE

When using CA the **mechanical index provides little dosimetric guidance for the safe use of CA** aside from a general indication of exposure risk. Such risk therefore needs to be mitigated where possible.

Strategies for reducing risk include:

- For normal and CEUS imaging keep MI to less than 0.4. For MI > 0.7 there *is* a risk of cavitation with CA.
- Follow the CA manufacturers guidance and warnings on use and dosage of CA
- Unless required for clinical investigation, avoid exposure of tissue that is particularly sensitive to cavitational damage, e.g. kidney.

QUALITY ASSURANCE AND ROUTINE CHECKS

Quality assurance (QA) is the name given to the practice of performing a set of routine checks to ensure that the ultrasound machine is performing to its design specification over the lifetime of the machine.

The importance of adequate QA is seen if one considers the problem of a reduction in quality.

- Poor images – loss of diagnostic information.
- Tendency to increase transmit power, lengthen examinations, etc. to compensate.
- Bio-safety – for example, probe damage resulting from damage during needle procedures or unclean probes.

For major QA checks and assessing problems, local medical physics or electronics departments, or the manufacturers of the scanner should be consulted. In particular they will be able to check that all is as it should be on acceptance of a new machine, when there are major upgrades and annually thereafter. Records of QA measurements should be recorded and kept. For more routine checks, the user needs to be responsible. They see the machine every day or week and they will be the first to notice any change in quality [4].

SUGGESTED ROUTINE USER CHECKS

Every Scan   Hand and probe hygiene is vitally important, especially when using endo-probes. Ensure probe and keyboard remain clean before and after each scan.

Every Day   Visually inspect probes for wear and tear and mechanical damage including needle damage on the face of the transducer. Check the monitor screen is clean and is correctly set up for brightness and contrast in normal viewing conditions.

Check the **air reverberation** pattern for any shadows or streaks in the image of each probe you use. This is simply an ultrasound image

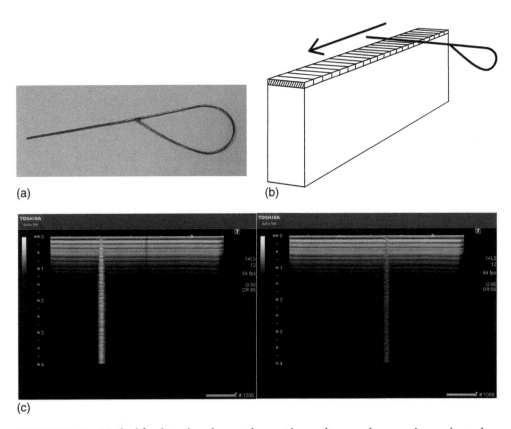

(a)   (b)

(c)

FIGURE 11.16   Method for detecting element dropout in an ultrasound array using a wire tool (a) and (b). An example showing the reduction in amplitude of the reverberation as the tool passes a non-functioning element (c). *Source:* Image Courtesy of Barry Ward.

of the probe in air with no gel and may indicate element drop-out and changes in the integrity of the front face of the probe (see below).

Weekly   Visually inspect probes, scanner controls and cables and brake function for wear and tear and mechanical damage.

Check for **element dropout** for each probe (Figure 11.16). This is a significant cause of image quality deterioration. It can be simply checked for, by running a thin metal wire tool (1 mm diameter), that can be made for the job, along the dry surface of the probe (see Figure 11.16b). The image of the wire should be uniformly seen across the probe and dropouts will be apparent as darker lines or absent echoes on the image.

Monthly   Check and clean air filters as needed.

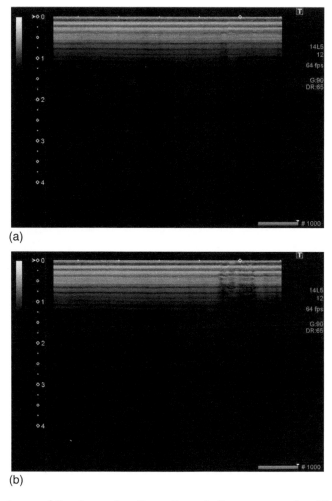

(a)

(b)

FIGURE 11.17   Image of the air reverberation pattern of a linear array probe with images (a) and (b) taken three months apart showing deterioration in the laminate on the front of the probe. *Source:* Images Courtesy of Barry Ward.

With **standard control settings**, use an **air reverberation test,** as described above, to check the sensitivity of each probe and **record the image**. The image should show a set of uniformly bright reverberation lines parallel to the front face of the probe. By recording the image, changes over time can be compared to spot changes that may indicate deterioration in image quality as shown in Figure 11.17.

Annual Service   Local medical physics or electronic department, or manufacturer.

THE USE OF TEST OBJECTS

More in-depth QA checks may be performed using test objects and in-vitro phantoms. These often consist of sets of wires or tubes of different diameters arranged in sets of patterns that allow axial and lateral resolution to be measured and measurement caliper calibration to be checked. They may be filled with a water-alcohol mixture to give a speed of sound of 1540 ms^{-1}. These may be **open top test objects (OTTO)** enabling the transducer to be positioned in the fluid (Figure 11.18). Alternatively, they may be filled with a tissue mimicking gel. The gel has scatterers embedded to mimic

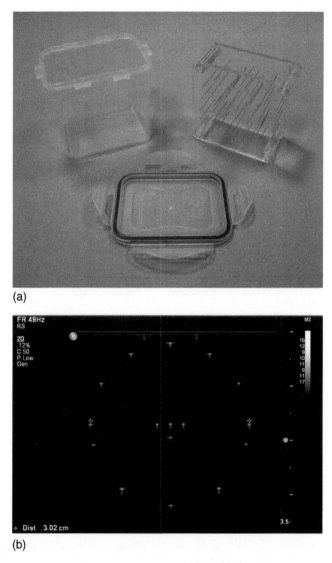

(a)

(b)

FIGURE 11.18　(a) Illustration of an open top test object (OTTO) consisting of wires accurately spaced in 1540 ms^{-1} fluid. (b) An image from the OTTO with measurement calipers shown.

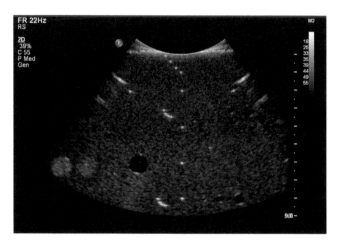

FIGURE 11.19 Image from a tissue mimicking phantom showing wires, cysts and solid lesions of different sizes and echogenicity.

the acoustic appearance of soft tissue. Tissue mimicking phantoms often have cylinders of higher and lower echogenicity and of varying diameter embedded in a mean echogenicity gel to allow contrast resolution and the detection of small lesions to be assessed as shown in Figure 11.19.

PERSONAL RISK MANAGEMENT

As well as ensuring safe practices regarding the use of ultrasound on the patient are maintained, it is also important to ensure that risks to you as a sonographer are properly managed. **Work-related repetitive strain injury (WRRSI)** is a potential hazard the sonographer faces [5]. The problem arises from the fact that scanning often involves micro-manipulation of the probe with the arm extended whilst at the same time turning to look at the screen and stretching the other arm out to operate the controls. For some types of scan, for example, vascular scans, TV scans and cardiac scans, the operator may need to adopt an awkward position in order to obtain the images needed. The problem with manipulating the probe with an extended arm is that all the muscles of the arm are in tension in order to produce a stable hand, that must then make very fine, well controlled, movements to visualise the various image planes.

In addition to operating the machine, the sonographer may have to perform some patient handling.

WRRSI Risk Mitigation

- Ensure couch height, machine height and chair height are all adjusted for comfort and ease of scanning – and try changing these for some variation.
- Position the screen so it is viewable with as little head twist or tilt as possible.

- Ensure the positioning of the patient is optimal for the scan being performed. All patient handling should be in accordance with good practice.
- Where possible rest your scanning arm whilst scanning, for example, on the patient where appropriate, on the couch or on a chair arm rest, to ensure the arm is as relaxed as possible. Frequently assess your posture for bad habits.
- Ensure you are holding the probe with an ergonomically efficient grip that does not stress or strain muscles and joints [6].
- Ensure background lighting in the room is appropriate without any reflected flare on the image screen (or flare across glasses if you wear them).
- During a scan take **micro-breaks** – that is, for a couple of seconds between views, consciously relax your arm and shoulders. Take your hand off the keyboard when not operating the controls.
- Between patients consciously perform stretch and relax exercises of your arm, shoulder, neck, etc.
- Manage your workload. Too many patients in a session increases the risks. Time between patients to write the report can be a time when a different set of muscles are used – ensure desk height, screen height and chair height for typing reports is appropriate.
- To use different muscle groups try scanning left-handed (or vice versa).

In Summary

- Find what works for you.
- Keep exercised and relaxed.
- Keep it under review.
- Share ideas with colleagues.
- With good risk management, problems should not occur and any risk be mitigated.

NEW TECHNIQUES IN ULTRASOUND

Finally, when considering quality related to the use of ultrasound, it is easy to overlook the question of the techniques and measurements we are making on patients. When a new modality or technical improvement in ultrasound is introduced, such as adaptive filtering or harmonic imaging, does it actually enhance the imaging we perform? Does it improve or at least maintain the quality of our work in using ultrasound diagnostically? This is perhaps an area of quality that is not often considered and it is easy to accept changes because they are presented to us by manufacturers. All of the techniques described in this book have been introduced, or are being introduced, as improvements. It is something to think about. One attempt to assess a new technique in a formal way is described in Hemmsen, M.C. et al. [7]. It uses the introduction of synthetic aperture imaging, described here in Chapter 13, as an example of moving from concept through to clinical assessment to ensure clinical value of the end product. Another looks at side by side comparisons of new beamforming and adaptive filtering techniques [8].

REFERENCES

1. Ter Haar, G. (ed.) (2012). *The Safe Use of Ultrasound in Medical Diagnosis*, 3e. London: British Institute of Radiology.

2. IEC 60601-2-37:2007+AMD1:2015 CSV (2015). Medical electrical equipment – Part 2-37: Particular requirements for the basic safety and essential performance of ultrasonic medical diagnostic and monitoring equipment. https://webstore.iec.ch/publication/22634 (accessed August 2022).

3. Jalal, J. and Leong, T.S.H. (2018). Microstreaming and its role in applications: a mini-review. *Fluids* 3: 93. https://doi.org/10.3390/fluids3040093.

4. Dudley, N., Russell, S., Ward, B., and Hoskins, P. (2014). BMUS guidelines for the regular quality assurance testing of ultrasound scanners by sonographers: BMUS QA working party. *Ultrasound* 22: 8–14. https://doi.org/10.1177/1742271X13511805.

5. Murphey, S. (2017). Work related musculoskeletal disorders in sonography. *Journal of Diagnostic Medical Sonography* 33: 356–369. https://doi.org/10.1177/8756479317726767.

6. Roll, S.C., Selhorst, B.S., and Evans, K.D. (2014). Contribution of positioning to work-related musculoskeletal discomfort in diagnostic medical sonographers. *Work* 47: 253–260.

7. Hemmsen, M.C., Lange, T., Brandt, A.H. et al. (2017). A methodology for anatomic ultrasound image diagnostic quality assessment. *IEEE Transactions on Ultrasonics, Ferroelectrics, and Frequency Control* 64: 206–217. https://doi.org/10.1109/TUFFC.2016.2639071.

8. Hasegawa, H. (2021). Advances in ultrasonography: image formation and quality assessment. *Journal of Medical Ultrasonics* 48: 377–389. https://doi.org/10.1007/s10396-021-01140-z.

CHAPTER 12

Advanced Topics

This chapter covers the following advanced topics:

- Contrast Agents
- B-flow Blood Vessel Imaging
- Doppler Measurement of Pressure Gradients
- Advanced Image Processing – Computer Aided Diagnosis
- Fusion Imaging
- Needle Visualisation and Guidance

CONTRAST AGENTS (CA)

For **contrast enhanced ultrasound (CEUS)** an agent is introduced into the bloodstream that will show up in contrast to the average B-mode speckle grey level. Blood normally shows as black or has very low-level echoes as it is a very poor echogenic target. Gas in the form of bubbles forms a highly echogenic target which when injected intravenously enables the blood space to be visualised right down to the micro-vessel level.

In its simplest form, gas bubbles may be introduced by simply injecting intravenous saline that has been shaken up. This is sufficient to demonstrate a patent defect in the heart as the strong signal of microscopic bubbles in the saline may be detected in the left carotid artery.

Ultrasound Technology for Clinical Practitioners, First Edition. Crispian Oates.

For contrast studies in general, a contrast medium of encapsulated gas micro-bubbles suspended in a carrier fluid is used [1]. The encapsulation stops the bubbles dissolving and prevents agglomeration of the bubbles.

The structure of a contrast bubble is an encapsulating shell enclosing a gas. Examples of the materials used are shown below.

Gas
- air
- perfluorocarbon
- nitrogen

Shell
- albumin
- galactose
- lipid polymers

NOTES

- Micro-bubbles are produced with a fairly uniform diameter of 1–4 µm. This may be compared with red blood cells ~8 × 2 µm (see Figures 12.1 and 12.2).
- Micro-bubbles are small enough to cross the capillary bed, including the lung, so remain in full circulation. They may last three to eight minutes.
- Heavier gases such as perfluorocarbon are less soluble so last longer in the circulation.
- Circulation around the body takes about one minute with the body at rest. Bubble loss occurs through filtration by the lungs, destruction by the immune system, passive dissolution, and shear forces in the circulation.

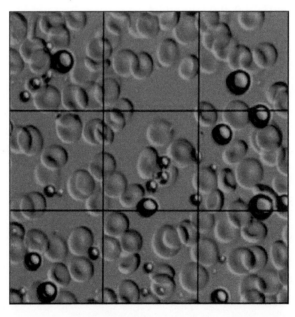

FIGURE 12.1   Perfluropropane bubbles with a protein shell (Optison; GE Healthcare, Milwaukee, Wis), seen here against a background of red blood cells *Source:* Wilson and Burns [2]/Radiological Society of North America.

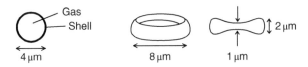

FIGURE 12.2   Comparison of size of a typical contrast microbubble and a red blood cell drawn to the same scale.

- Micro-bubbles are large enough to not cross the vascular endothelium. This means they stay within the vascular space – unlike CT or MRI contrast agents which quickly migrate out of the vascular space.
- Leaks from the vascular space may therefore be detected, for example endovascular leaks in aortic stents.
- As bubbles stay within the vascular space and they are a strongly reflecting agent, typically, only 1–2 mL of contrast is needed.

BEHAVIOUR OF BUBBLES IN THE ULTRASOUND FIELD

Bubbles, including the encapsulated bubbles used as contrast agents, will oscillate in the sound field as the acoustic pressure changes. At low pressure, they expand and at higher pressure they are compressed (Figure 12.3).

The relative change in diameter will depend on the peak-to-peak pressure change in the sound wave and the stiffness of the bubble shell.

- The more elastic the shell wall, the greater the diameter excursion.
- Ambient pressure change such as systolic-diastolic BP, will affect bubble diameter.

As bubbles oscillate in the sound field, they are not simply passive reflectors of the ultrasound wave but they re-radiate the sound wave as acoustic sources themselves.

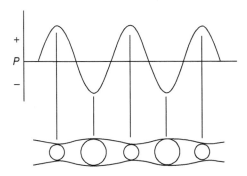

FIGURE 12.3   Change in microbubble diameter with pressure change in the ultrasound wave.

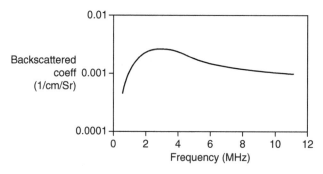

FIGURE 12.4   Backscattered ultrasound signal vs frequency for the experimental contrast agent BR14 (Bracco Research SA, Geneva, Switzerland). The wide peak around 3 MHz reflects the fact there is a range of bubble sizes present in the sample. *Source:* Adapted from de Jong et al. [3].

The strength of the echo signal depends on the magnitude of the oscillation of the bubbles. This will be greatest when the transmitted ultrasound frequency matches the **resonant frequency** of the bubbles. This is similar to pushing a child on a swing. When the 'push' occurs at the natural frequency of the swinging child (pendulum), the amplitude of the swing increases. The driving force and the swing are then **resonant**.

It is a convenient property of the bubble size used in contrast media that they resonate at the megahertz frequencies used for diagnostic ultrasound (Figure 12.4). The transmit frequency can then be tuned to enhance the echoes received from the contrast.

CONTRAST AGENT HARMONICS

When insonated with low intensity ultrasound, they resonate linearly with the transmitted ultrasound frequency. At higher intensities, their mechanical oscillation becomes **non-linear** as it is easier for bubbles to expand than to contract. As a consequence of this non-linear oscillation, they radiate **harmonics** of the incident ultrasound frequency. The echo received from the bubbles contains second and third harmonics (see Chapter 5) and a subharmonic at half the transmitted frequency. When encapsulated bubbles burst, they release their gas as free bubbles. The gas dissolves very quickly, but as it does so the bubbles radiate harmonics very strongly from the insonating pulses.

The example in Figure 12.5 shows the harmonics generated for a low intensity 1 MHz ultrasound beam 0.3 MPa and a higher intensity beam 0.6 MPa. At low intensity, the scattered sound is mainly at the insonating frequency, but a subharmonic at half the transmitted frequency and second and third harmonics are also present. At higher intensities, a series of harmonics are generated in the backscattered sound.

Contrast agents will show up on a normal B-mode image, but the **contrast ratio** between the contrast agent and the tissue B-mode image can be significantly enhanced by using techniques to detect the harmonics arising from the contrast [4].

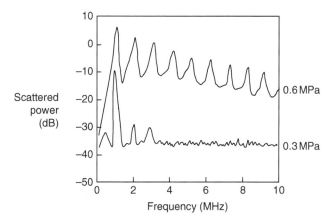

FIGURE 12.5 Backscattered sound from a 1 MHz transducer insonating Quantison contrast agent at two transmitted powers. The trace at 0.3 MPa shows a subharmonic at half the transmitted frequency as well as second and third harmonic peaks. *Source:* Adapted from de Jong et al. [3].

Amplitude Modulation

The first technique uses a sequence of two pulses, where the first pulse T_1 has double the amplitude of the second T_2, enables the non-linear signal from bubbles to be detected as shown in Figure 12.6.

On receive, the echoes from the second, half amplitude, pulse T_2 are amplified by a factor of two and then subtracted from the echoes from the first, larger, pulse T_1. Where the pulses have behaved in a linear way in the tissue, the two echoes subtract to give a net zero signal. However, scattered echoes arriving from the bubbles will be different for the two different amplitude pulses due to their non-linear behaviour. Echoes arriving from contrast will therefore have a non-zero signal when the two pulses are subtracted. By this means, the contrast will be detected and its signal separated from that of the surrounding tissue.

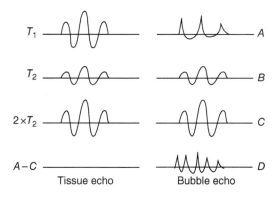

FIGURE 12.6 Amplitude modulation, where T_2 is the same as T_1 but half the amplitude, is used to remove background echoes (a) and allow harmonic echoes from the contrast agent to be displayed (b). The echoes from T_2 are doubled to give C and C is added to A, to give D.

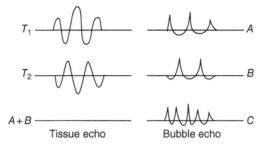

FIGURE 12.7   Pulse inversion, where T_2 is the same as T_1 but with the phase inverted, is used to remove background echoes (a) and allow harmonic echoes from the contrast agent to be displayed (b). The echoes from T_1 and T_2 are added to give C.

Pulse Inversion

The second way of detecting contrast bubbles is to use the pulse inversion technique (Figure 12.7).

For this, two similar pulses are transmitted T_1 and T_2 but for the second T_2, the phase is inverted. Subtracting the echo of the inverted transmit pulse T_2 from that of the non-inverted pulse T_1 removes the tissue signal but leaves the harmonic contrast signal.

Both of these techniques separate the non-linear harmonic echoes from the tissue echoes. The contrast signal may then be displayed together with the B-mode tissue image or, in a similar manner to X-ray subtraction angiography, a **contrast only image** may be produced (Figure 12.8). The techniques can be combined and by using

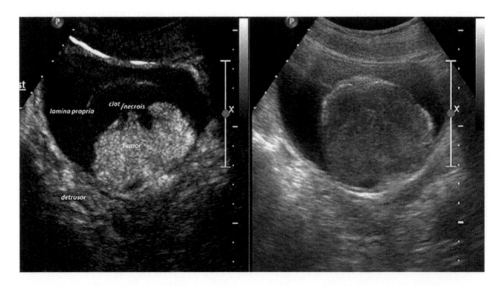

FIGURE 12.8   A harmonic contrast image of a urothelial carcinoma in the bladder with normal B-mode image shown for comparison. *Source:* Gupta et al. [5]/Central European Journal of Urology.

contrast pulse sequencing tissue movement suppression, further enhancement of contrast can be achieved. Contrast pulse sequencing is the technique of interrogating a scan line a number of times using a sequence of different phases and amplitudes and then combining the echo signals [4].

Ultrasound contrast media may be administered in one of two ways:

- Using a bolus injection and watching perfusion as the contrast fills a vascular space.
- Using a continuous injection so the circulation fills with contrast.

FLASHING

Using a continuous injection opens up the possibility of using contrast for quantitative measurements. By applying a single large amplitude transmit pulse with a large P_- (negative pressure), known as **flashing**, the bubbles can be made to burst, with their gas dissolving in the bloodstream. Following removal of contrast, by flashing the bubbles in this way, quantitative measurements of contrast intensity increase during refill from the circulating contrast agent can be made [6]. This can give useful information on organ perfusion. Such measurements require careful, reproducible adjustments of the scanner controls, in particular output power, focal depth, dynamic range, receive gain and transmit frequency.

NOTES

- Unlike X-ray or MRI contrast techniques, ultrasound contrast is unique in being able to be removed and then re-gained in this way.
- The technique of flashing to destroy micro-bubbles followed by reperfusion measurements is particularly suited to the ultrafast ultrasound technique where low amplitude plane waves preserve the micro-bubbles and measurements at high frame rates can be made (see Chapter 14).

ADVANCED MICRO-BUBBLE TECHNIQUES

Micro-bubbles may be used **diagnostically** and **therapeutically** by labelling the micro-bubble shell so that it binds to a particular cell receptor as shown in Figure 12.9. It may then be used to carry a therapeutic drug or vector DNA to that site. The labelled bubbles will bind to the receptor site to highlight the site on the ultrasound image, for example, areas of inflammation, active thrombi or cancer cells [7].

For therapeutic use, the bubbles, having been delivered to the target of interest, can be flashed with a large ultrasound pulse so they break and deliver their therapeutic agent directly to the target.

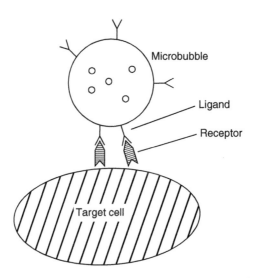

FIGURE 12.9   Showing how microbubbles may be labelled so as to attach themselves to receptors on specific targets.

NOTES

- Be aware of the safety issues involved in using contrast agents as discussed in Chapter 11.
- Contrast agents should always be used in accordance with the manufacturer's advice.

B-FLOW BLOOD VESSEL IMAGING

This technique enables blood flow to be shown on B-mode images with a similar greyscale appearance to the B-mode tissue image (Figure 12.10).

Echoes from blood form a similar speckle pattern to those of the surrounding tissue. However, blood is a very poor echogenic target with an echo strength −60 dB below that of surrounding tissue. On a B-mode image, fast moving blood shows up as black.

At the low velocities seen in veins and in the cardiac chambers, some echoes are seen from blood as the red blood cells then stack up to form what are called **rouleaux,** which form larger echo targets (Figure 12.11). This is the cause of the so called '**whisp of smoke**' sign as seen in Figure 12.12.

One way to enhance the signal from blood would be to increase the transmit power but this would also increase the tissue signal by a similar amount and would have safety implications. Therefore, in order to show the echoes from blood at a level that may be compared with other tissues, it is necessary to isolate the blood signal and to amplify it relative to the surrounding tissue signal. B-flow achieves this, enabling blood to be seen on a B-mode image.

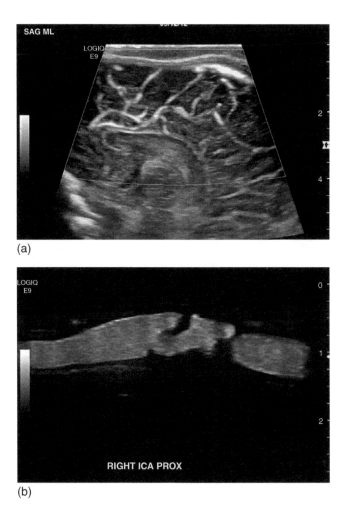

(a)

(b)

FIGURE 12.10 B-flow images showing (a) blood flow in the neonatal brain and (b) a B-flow only image of the carotid artery with low echogenic plaques [8].

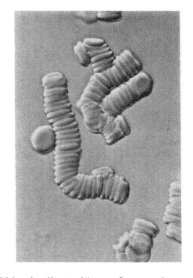

FIGURE 12.11 Image of red blood cells stacking to form rouleaux.

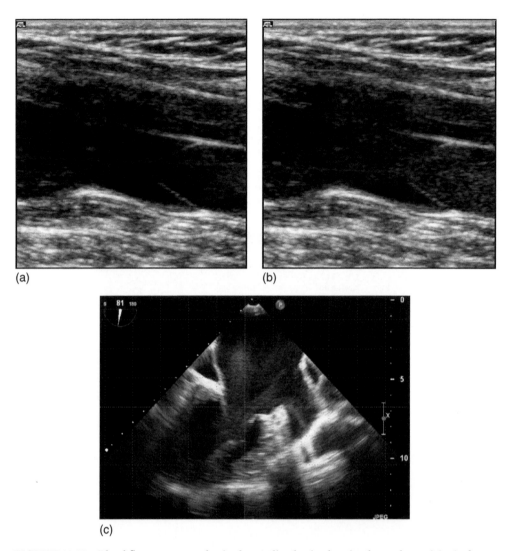

(a)   (b)

(c)

FIGURE 12.12   Blood flow across a valve in the popliteal vein showing low echogenicity in fast flow (a) and increased echogenicity in stationary flow (b). (c) Shows 'Whisp of smoke' in a dilated left atrium. *Source:* Courtesy of Chris Eggett.

The process of isolating the blood signal uses **coded excitation** to detect what is moving in the field of view. The coded pulses are detected and compressed to give a strong signal from weak targets, equivalent to a short high amplitude pulse (see Chapter 6).

B-flow uses coded excitation to detect movement in the field of view. The principle of how the separation of blood flow signal may be achieved is as follows. Using a sequence of two transmitted pulses (P_1, P_2), slowly moving blood may be detected by comparing the speckle pattern from a small sample volume within the ultrasound beam. The pairs of pulses are compared using cross-correlation (see Chapter 9).

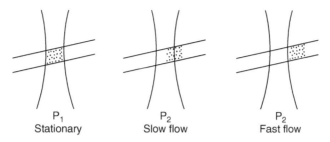

FIGURE 12.13 Showing the change in speckle between two successive pulses P_1 and P_2 seen by an ultrasound beam with stationary, slow, and fast flow.

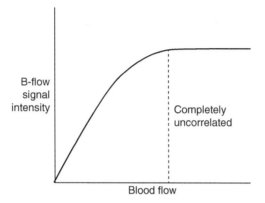

FIGURE 12.14 Relationship between the B-flow signal and the correlation of the speckle as flow velocity increases.

The difference between P_1 and P_2 will be zero where the tissue has remained stationary but there will be a net signal where the speckle pattern has changed, i.e. moving blood. The faster the blood moves, the greater the net blood signal detected, up to a maximum value that will occur when all of the blood seen by the first pulse has left the sample volume by the time the second pulse arrives as shown in Figures 12.13 and 12.14.

The strength of the blood signal is then displayed within the vessels on a B-mode image.

NOTES

- B-flow enables the visualisation of blood flow on a B-mode image.
- The signal strength depends on the speed and turbulence of the blood flow. Turbulent flow decorrelates faster than laminar flow.
- Small – sub-millimetre vessels may be detected.
- As no wall filter is required, the B-flow image shows flow right up to the vessel wall.

- The level of B-flow brightness to B-mode image brightness can be adjusted to improve contrast or remove the B-mode altogether so only blood flow is shown.
- The B-flow signal may be colour coded to increase its contrast against the B-mode image.

DOPPLER MEASUREMENT OF PRESSURE GRADIENTS

It is useful to be able to measure the pressure drop ΔP across a vascular stenosis or a stenosed heart valve as this indicates the haemodynamic significance of the stenosis [9]. Doppler ultrasound may be used to estimate the pressure drop by measuring the velocity of flow proximal to the stenosis v_p at peak systole and the highest velocity found in the stenotic jet v_s [10].

Using a simplified version of the **Bernoulli equation**, the pressure drop is calculated as follows:

$$P_s - P_p = \frac{1}{2}\rho\left(v_s^2 - v_p^2\right) = \Delta P \text{ units Pa}\left(\text{pascal}\right)\text{or kg·m}^{-1}\text{s}^{-2}$$

where ρ is the density of blood $= 1060 \text{ kg·m}^{-1}$.

Applying a conversion factor of $1 \text{ mmHg} = 133.3 \text{ Pa}$, the equation can be written

$$4\left(v_s^2 - v_p^2\right) = \Delta P \quad \text{in mmHg}$$

where $v_s \gg v_p$, as is often the case when comparing the velocity in a cardiac stenosis with the velocities in the cardiac chamber, the equation can be further simplified to:

$$\Delta P = 4v_s^2$$

An example of pressure measurement across an aortic stenosis is seen in Figure 12.15.

Where velocity vector data is available for the whole flow across the stenosis, for either 2D or 3D data, then the pressure drop may be more accurately estimated using the **Navier-Stokes equation**, which gives a full description of the flow in 3D [11].

$$\nabla P = -\Delta\left(\frac{\partial \mathbf{v}}{\partial t} + \mathbf{v} \cdot \nabla \mathbf{v}\right)$$

where ∇P is the change in pressure P along a given vector, $\dfrac{\partial \mathbf{v}}{\partial t}$ is the change in velocity with time along the vector, $\mathbf{v}$ is the vector at a point and $\Delta \mathbf{v}$ is the change in velocity along the vector (see Chapter 13 for velocity vectors). These calculations are performed by programmed algorithms in a velocity vector processing package.

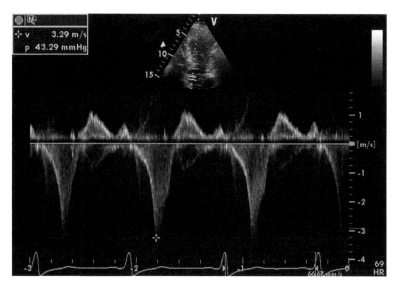

FIGURE 12.15 High velocities in the left ventricular outflow tract with the pressure drop across the stenosis indicated. Multiple sample gates are being used to increase the Nyquist limit. *Source:* Courtesy of Chris Eggett.

ADVANCED IMAGE PROCESSING

Having obtained some image data there are ways in which the data may be processed to:

- Make what is seen easier to interpret.
- Make automated measurements from the data.

Both of these are being used on ultrasound images. In order to achieve these ends, the scanner must identify key features in an image. The advent of 3D ultrasound has enabled the range of what can be done to be greatly extended.

ARTIFICIAL INTELLIGENCE

There are a number of steps involved in such image processing and sophisticated algorithms are used to automate the process. These methods fall under the general heading of **artificial intelligence (AI)**. Artificial Intelligence can be broadly defined as any human like behaviour displayed by a machine or system. **Machine learning** is the study of algorithms that can improve automatically through experience and by the use of data. One subset of machine learning techniques that has proved very effective in enabling automatic analysis of images and other AI processes is called **deep learning** and is discussed further below.

SEGMENTATION

Image segmentation is the process of identifying edges, surfaces, and textures visible in an image so that the image is divided into regions of interest. This may be achieved by manually identifying a region or structure, for example by using a cursor, or it may be automated. Automated methods may require the user to first identify a general region or **segmentation seed** in the image on which the automatic technique will work.

At its most basic, it may be as simple as using the data from a particular modality, such as CDU, and displaying blood flow in a more informative way. Or, detecting the first interface in a clear fluid path, such as detecting the foetal skin surface within the amniotic sac. This can then be used to produce advanced 3D displays of the foetus in utero.

EXAMPLES (1–3)

1. **Displaying Blood Vessels**
   The lumen of a blood vessel may be detected by looking for the surface between the blood of low echogenicity and the strongly reflecting vessel wall. An alternative approach is to use Doppler information or the signal from CEUS to identify the space where there is flow (Figure 12.16).

2. **Displaying the Foetus**
   The skin surface of a foetus in utero may be identified by using first surface detection of echoes within the liquor in the amniotic sac. Once a surface has been segmented it can be highlighted in the displayed image.

   Using methods developed for video games, a point of illumination can be chosen and the surface rendered so as to appear the same as if you saw it in front of you, with a light casting shadows to show facial features (Figure 12.17). Internal organs may be displayed by making the surface semi-transparent.

3. **Cardiac Valves**
   In a similar manner to that used for displaying the foetus, cardiac valves may be rendered to show what one would see during surgery (Figure 12.18). This allows planning for prosthetic valve insertion. Using 4D data the motion of the valve may also be viewed.

   Flow can be superimposed to show where a regurgitant jet is emanating from, as shown in Figure 12.19.

At the other end of the scale, segmentation may be used to detect solid lesion boundaries within tissue parenchyma, identify image orientation, or identify the boundaries and fit a model to the volumes identified. In the case of echocardiographic imaging, this can involve identifying frames in a cine sequence and tracking changes

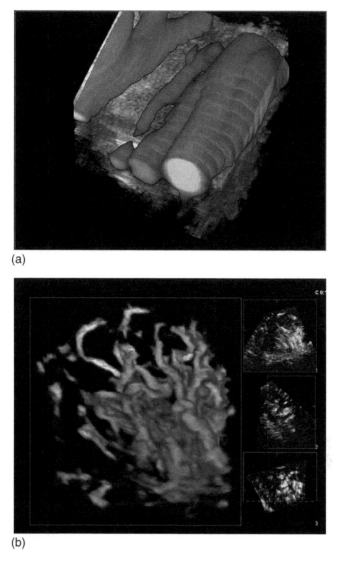

(a)

(b)

FIGURE 12.16 (a) 3D rendering of the carotid bifurcation using the power Doppler signal. *Source:* Courtesy of Esaote. (b) 3D rendering of the vasculature of a hepatocellular carcinoma revealed by CEUS. *Source:* Illustration by Courtesy of Koninklijke Philips N.V.

across the cardiac cycle. From basic segmentation, we move into the realm of computer aided diagnosis.

COMPUTER-AIDED DIAGNOSIS (CAD)

CAD may be used to identify key images for human analysis/detection or may go further to automatically provide a diagnosis [13].

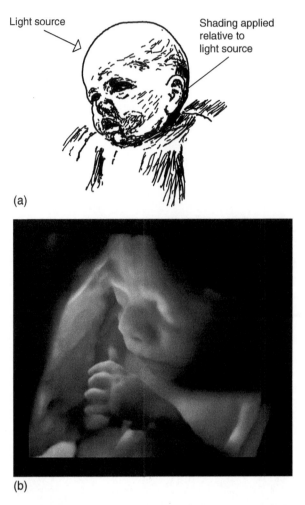

(a)

(b)

FIGURE 12.17 Foetal surface shown using a 'light source' to give a 3D rendering. *Source:* Illustration by Courtesy of Koninklijke Philips N.V.

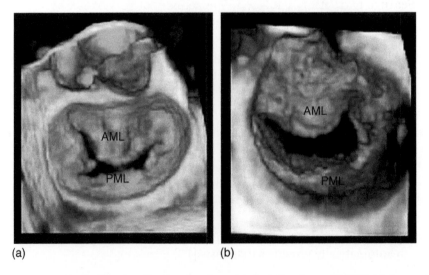

(a)   (b)

FIGURE 12.18 Mitral valve showing the view from (a) the left atrium and (b) the left ventricle. *Source:* Drasutiene et al. [12].

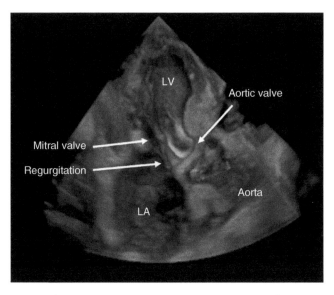

FIGURE 12.19   Mitral valve regurgitation shown on a 3D rendering of the heart.
Source: Illustration by Courtesy of GE Healthcare.

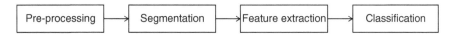

FIGURE 12.20   Block diagram of the process of computer aided diagnosis.

The method requires the following steps (Figure 12.20):

- **Pre-processing** – to ensure the data is in the right form for analysis.
- **Segmentation** – to isolate the regions for analysis.
- **Feature extraction** – to identify the features of significance and present relevant measurements.
- **Classification** – use measurements, or otherwise determine the significance of the features to categorise the information and possibly offer a concluding result that may be a clinical result.

Segmentation and feature extraction may be closely linked in that the techniques used to identify what is significant may also be the same as those used to segment the image. Feature extraction may go further than basic segmentation and highlight features within the segmented areas upon which classification can be made [14].

Having segmented the data to isolate the relevant image region, the next step is to see if it has the features that indicate a significant finding. This has been done in two ways:

- Direct identification of features.
- Using deep learning to identify features.

Feature Identification

Four feature categories have been used to identify features of interest:

- **Texture** – e.g. character of lesion surface; local greyscale texture.
- **Morphology** – shape of lesion, e.g. smoothness of lesion margin; depth to width ratio, etc.
- **Model-based** – e.g. statistical analysis of backscattered echoes.
- **Descriptor features** – direct description of what is seen in the image, e.g. shape (round, oval, irregular); calcification; shadowing, etc.

Examples

1. **Cardiac Quantification**
   Using a pre-defined model, the segments of the model heart are matched to a 3D volume set of images of a real heart in a series of adjustments that, in four stages, give a good match (Figure 12.21). The model is fitted across the cardiac cycle. Using the fitted model calculations of the chamber volumes and other cardiac functional parameters may be calculated. The modelling and fitting use an algorithm derived using deep learning. An alternative approach aims to model the left ventricle directly from a set of ultrasound images [15], again using a deep learning algorithm.
2. **Soft Tissue Segmentation**
   Using a deep learning algorithm, the segmentation of a soft tissue mass, where the change in speckle texture is subtle, can be accurately achieved. In this case for a benign thyroid nodule with degenerative changes (Figure 12.22).

Classification

Having identified the relevant features, the target is **classified**. For example, as benign or malignant. This may be done in a number of ways, for example, using a **Bayesian statistical** approach (probability of positive) or using a **decision tree** based on the features found, etc.

Deep learning has been applied in a number of stages in the CAD process, for example in the models used to fit to cardiac images for segmentation, in feature extraction and in classification [17, 18].

Deep Learning

This uses **artificial neural networks** to learn to recognise the features of interest in an image. A neural network is an arrangement of computations that in some ways mimic the way neurons in the brain work. The computations are arranged in layers such that the outputs from one layer provide the inputs to another layer. The outputs

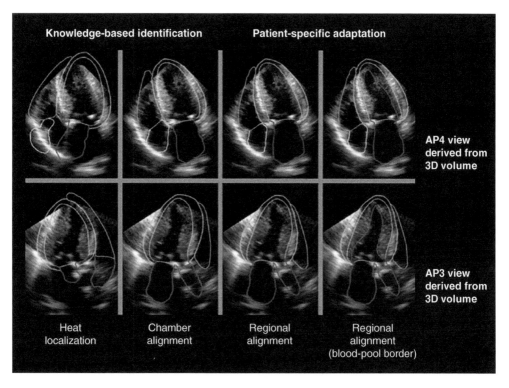

FIGURE 12.21 Automated segmentation of the heart using four stages as shown across the bottom of the image for two views of the heart (top and bottom lines). Observe how the model is gradually fitted from the left to the right in each line. *Source:* Illustration by Courtesy of Koninklijke Philips N.V.

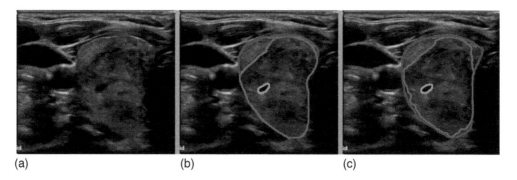

FIGURE 12.22 Segmentation of a benign thyroid nodule with degenerative changes using a deep learning algorithm. (a) the ultrasound image, (b) manual segmentation (normal thyroid blue, nodule red, cyst green) and (c) automated segmentation using the algorithm. *Source:* Kumar et al. [16]/IEEE/CC BY 4.0.

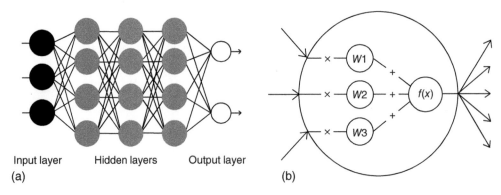

FIGURE 12.23   Schematic of a neural network (a). Each node receives input from every node in the previous layer (b). The inputs are multiplied by a weighting factor and then summed and processed by an activation function to give an output fed to each node on the next layer. *Source:* Adapted from Fortunati [19].

from each layer are modified by the status of the inputs to the layer in an analogous way to neuron outputs being modified by input signals (Figure 12.23). In this way the system is able to 'learn' and produce useful outputs for future input data.

The deep learning method requires a large dataset of known cases to train a **neural network**. The example cases are divided into two. One half is used to train the network with normal and abnormal images, for example. These are known as **ground truth data**. The network then automatically extracts the features it determines, to classify the images. These features may be completely different to any that a human might suggest, such as those described above. It is to this extent like a **black box** – you don't know what goes on inside but the answer comes out of the output. Having trained the network, the second half of the examples are then used to test the ability of the network to correctly classify the data.

DIAGNOSIS WITH CAD

Having performed the automated analysis, the user is presented with some form of result. This may be in the form of a categorical result (e.g. yes/no) or a probabilistic result (e.g. a scale or indication of probability). With all automated diagnostic methods there is the question of robustness, reliability and the potential for false results. The same is true for a human diagnostician arriving at a result, but where a machine derived result is presented to you, the onus is on you to assess whether the result can be relied upon. Firstly, in the immediate situation, for example, has the correct area in the image been highlighted, does it look likely. And secondly, is there data to support the particular CAD algorithm's results in that area of examination. How does it correspond with your own diagnostic opinion of what is seen? As automated analysis becomes more prevalent, the risk of divorcing active analysis of the ultrasound data from simply accepting the machine derived result is a possible issue.

Examples of CAD

CAD has been used in ultrasound to classify breast lesions, liver fibrosis, and thyroid nodules. Accuracies are in the range 90–99% and in some instances may outperform human classification from ultrasound [20–22].

FUSION IMAGING

Fusion imaging is the technique of simultaneously using images of a target of interest from more than one imaging modality. In particular the use of **simultaneous synchronous navigation** of a 3D reference image set, together with live ultrasound scanning. Such a reference image set may be from CT, MRI, or PET scanning [23].

This enables the sonographer to more easily observe and carry out a procedure on a target that shows up clearly on a non-live image from another imaging modality whilst using a live ultrasound scan to direct the procedure, for example, aspiration, biopsy, or other minimally invasive procedure. It is especially useful for areas of poor imaging on ultrasound for example low contrast lesions, shadowed areas from gas or calcification or where there is poor image quality resulting from overlying fatty tissue.

The benefits of fusion imaging are:

- Increased confidence in identifying what is observed on the ultrasound image.
- Comparison of features as seen in different modalities.
- Saving on more expensive procedures such as surgery.
- Reducing false positives when treating small lesions under ultrasound guidance.
- For use as a training tool to understand the appearance of structures in different modalities.

In order to achieve fusion imaging three conditions are required:

- **Scaling** – The reference dataset and the ultrasound image must have the same scale so that dimensions on both images match when they are overlaid.
- **Registration** – The two image planes must be aligned in 3D space so that the two imaging modalities are showing exactly the same slice through the target of interest. This involves **re-slicing** the reference data set to match the ultrasound image plane.
- **Synchronous movement** – The displayed image plane of the reference data must continuously re-align to match the current image plane displayed by the ultrasound, in real time.

We therefore need to know the position and orientation of the ultrasound probe at all times in order to know the current image plane in space. This **tracking** is achieved

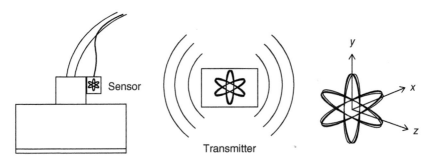

FIGURE 12.24   The position and orientation of the probe may be tracked using an electromagnetic sensor attached to the probe (a). This picks up pulsed magnetic fields from a transmitter mounted nearby (b). The transmitter and sensor have coils or other magnetic sensors mounted in three orthogonal planes to give full 3D information (c).

using an electromagnetic (EM) sensor attached to the ultrasound probe. A magnetic field is generated by currents passed through coils in a **field source** or **transmitter** situated less than half a metre from the patient. A small **position sensor** is attached to the ultrasound probe. When the probe is moved within the magnetic field, an induced electric current is generated allowing the system to recognise the 3D spatial **position** and **orientation** of the ultrasound probe (Figure 12.24).

Registration with the reference data is achieved either by using **external markers** placed on the patient's skin at the time the reference image data was collected or by aligning **anatomical markers** – vessels, cysts, calcifications, the xiphoid process, etc. In the case of anatomical markers, alignment is either performed manually or automatically using intelligent software. Once registration has been achieved, the two images can be locked together and synchronised in further positioning of the ultrasound probe. As the scan proceeds, registration may have to be readjusted to match the reference image to the ultrasound image for the exam to continue.

NOTES

- The fusion images of the two modalities may be displayed side-by-side or superimposed on one another.
- Ultrasound modes such as CDU, CEUS and elastography may be used to enhance contrast of the target of interest.
- The ideal track for inserting a biopsy needle, for example, may be planned in advance of the examination and shown on the images and the progress of the physical needle tracked and matched to the ideal track.

An example of fusion imaging is shown in Figure 12.25. There are a number of potential problems with fusion imaging:

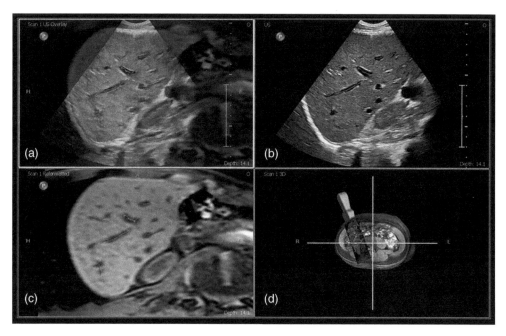

FIGURE 12.25   An example of fusion imaging of CT and ultrasound in the liver. (a) ultrasound superimposed on CT, (b) ultrasound, (c) CT and (d) indication of the probe orientation and slice on the CT. Registration was achieved automatically using anatomical landmarks. *Source:* Illustration by Courtesy of Koninklijke Philips N.V.

- Relative movement between the live image and the reference image reducing registration accuracy, for example due to breathing or probe pressure on the patient's skin.
- Lack of anatomical landmarks such as vessels within the field of view.
- Registration error developing as the scan proceeds.
- Interference of the magnetic fields from metallic objects, or placing the field generator too far away, causing inaccurate information from the sensor of the current ultrasound image plane.

NEEDLE VISUALISATION AND GUIDANCE

Real time ultrasound has a vital role in guiding needle procedures such as biopsy, aspiration, administering anaesthetic blocks, placing central lines, foetal surgery, etc. These require clear visualisation of the position of the needle tip and knowledge of its direction of travel.

When a needle is introduced into the image plane perpendicular to the ultrasound beams it will be seen as a bright reflection often with reverberation artefact, or comet sign, extending below it (see Figure 7.23). However, it is often not possible to achieve this alignment, especially when trying to reach deeper structures. In such cases the needle itself may be barely visible in the image. Larger needles show up better than fine needles and, if bevelled, the needle tip may scatter the ultrasound so as to show up on the image. A number of techniques have been used to improve visualisation of needles and enable them to be guided to their intended target.

Many needle procedures are performed freehand, but **needle guides** may be fitted as an attachment to the ultrasound probe (Figure 12.26). The line of the needle may then be displayed on the image as it is travelling along a known path. The target may be approached using a planned angle/depth route either in a longitudinal plane, in line with the long axis of the probe, or in a transverse, short axis direction. The needle guide ensures the planned angle is maintained.

The problem of knowing exactly where the needle tip is, is compounded by several possible problems.

- As the needle is pushed through the tissue there may be some flexing of the needle, resulting in the tip not being exactly on the planned line.
- The needle may appear bent due to refraction artefact (see Figure 7.8).
- The slice thickness of the image is not zero so within the longitudinal approach, the needle may still be seen in the image but be to one side of the target, for example, a small blood vessel viewed in longitudinal section. Rocking the probe along its long axis may make this situation clear (Figure 12.27).
- Using the transverse approach, it is easy to pass the needle tip through the image plane and beyond it. If this occurs, the acoustic shadow from the needle may be seen.

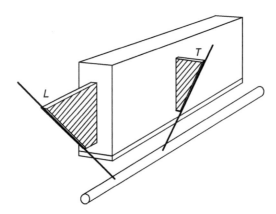

FIGURE 12.26   Diagram of a longitudinal L and transverse T needle guide on a linear array probe.

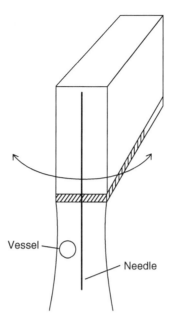

FIGURE 12.27   Illustrating the problem of the needle and target both being within the slice thickness but not coincident.

NOTES

- Tracking needle progress may be made in real-time on the image by noticing the disturbance to adjacent tissue structures as the needle pushes through.
- Needles are best seen in the focal zone, so this should be adjusted to ensure optimal visualisation.
- Larger needles are more easily seen than fine needles.
- Air bubbles in saline may be introduced to show where the needle tip is but these then have the disadvantage of obscuring the image at greater depth.

A number of more advanced techniques have been used to improve needle visualisation [24].

Making the needle **acoustically reflective**. This may be achieved either by coating the needle with an acoustically reflective coating or by laser etching the surface of the needle to make it scatter, rather than reflect, the ultrasound.

Active sensing, for example using an electromagnetic sensor or an ultrasound transmitter at the tip or an ultrasound emitting tip, to enable the tip position and track to be actively displayed on the ultrasound image [25]. This enables the needle to be homed in on the target of interest whilst avoiding other sensitive structures.

Image-based automatic needle tracking has been developed to continuously display its position in the image. This may be done by analysing changes in the B-mode image caused by the presence of the needle or by using power Doppler to detect needle vibrations.

Using such tracking methods, together with 3D imaging, the exact track of the needle in space may be visualised.

REFERENCES

1. Stride, E. (2015). Physical principles of microbubbles for ultrasound imaging and therapy. *Frontiers of Neurology and Neuroscience* 36: 11–22. https://doi.org/10.1159/00036622.

2. Wilson, S.R. and Burns, P.N. (2010). Microbubble-enhanced us in body imaging: what role? *Radiology* 257: 24–39.

3. de Jong, N., Emmer, M., van Wamel, A. et al. (2009). Ultrasonic characterization of ultrasound contrast agents. *Medical & Biological Engineering & Computing* 47: 861–873.

4. Whittingham, T.A. (2005). Contrast-specific imaging techniques: technical perspective. In: *Contrast Media in Ultrasonography, Basic Principles and Clinical Applications* (ed. E. Quaia), 43–70. Berlin: Springer.

5. Gupta, V.G., Kumar, S., Singh, S.K. et al. (2016). Contrast enhanced ultrasound in urothelial carcinoma of urinary bladder: an underutilized staging and grading modality. *Central European Journal of Urology* 69: 360–365.

6. Tang, M.X., Mulvana, H., Gauthier, T. et al. (2011). Quantitative contrast-enhanced ultrasound imaging: a review of sources of variability. *Interface Focus* 1: 520–539. https://doi.org/10.1098/rsfs.2011.0026.

7. Lee, H., Kim, H., Han, H. et al. (2017). Microbubbles used for contrast enhanced ultrasound and theragnosis: a review of principles to applications. *Biomedical Engineering Letters* 7: 59–69. https://doi.org/10.1007/s13534-017-0016-5.

8. GE White Paper (2012) B-flow Technology, Matt Berger, Qian Adams GE Healthcare 2012.

9. Nguyen, T., Hansen, K.L., Bechsgaard, T. et al. (2019). Non-invasive assessment of intravascular pressure gradients: a review of current and proposed novel methods. *Diagnostics* 9: 5. https://doi.org/10.3390/diagnostics9010005.

10. Harris, P. and Kuppurao, L. (2016). Quantitative Doppler echocardiography. *BJA Education* 16: 46–52. https://doi.org/10.1093/bjaceaccp/mkv015.

11. Olesen, J.B., Villagómez, H.C.A., Møller, N.D. et al. (2018). Non-invasive estimation of pressure changes using 2-D vector velocity ultrasound: an experimental study with in-vivo examples. *IEEE Transactions on Ultrasonics, Ferroelectrics, and Frequency Control* 65: 709–719. https://doi.org/10.1109/TUFFC.2018.2808328.

12. Drasutiene, A., Sigita Aidietiene, S., and Diana Zakarkaite, D. (2015). The role of real time three-dimensional transoesophageal echocardiography inacquired mitral valve disease. *Seminars in Cardiovascular Medicine* 21: 16–26. https://sciendo.com/article/10.2478/semcard-2015-0003.

13. Huang, Q., Zhang, F., Li, X. et al. (2018). Machine learning in ultrasound computer-aided diagnostic systems: a survey. *BioMed Research International* 5137904. https://doi.org/10.1155/2018/5137904.

14. Bass, V., Mateos, J., Rosado-Mendez, I.M. et al. (2021). Ultrasound image segmentation methods: a review. *AIP Conference Proceedings* 2348: 050018. https://doi.org/10.1063/5.0051110.

15. Kim, T., Hedayat, M., Vaitkus, V.V. et al. (2021). Automatic segmentation of the left ventricle in echocardiographic images using convolutional neural networks. *Quantitative Imaging in Medicine and Surgery* 11: 1763–1781. https://doi.org/10.21037/qims-20-745.

16. Kumar, V., Webb, J., Gregory, A. et al. (2020). Automated segmentation of thyroid nodule, gland, and cystic components from ultrasound images using deep learning. *IEEE Access* 8: https://doi.org/10.1109/access.2020.2982390.

17. Wang, Z. (2022). Deep learning in medical ultrasound image segmentation: a review. *arXiv:2002.07703v3 [eess.IV]* https://doi.org/10.48550/arXiv.2002.07703.

18. Looney, P., Stevenson, G.N., Nicolaides, K.H. et al. (2018). Fully automated, real-time 3D ultrasound segmentation to estimate first trimester placental volume using deep learning. *Journal of Clinical Investigation Insight* 3: e120178. https://doi.org/10.1172/jci.insight.120178.

19. Fortunati, V. (2019). Quantib. https://www.quantib.com/blog/how-does-deep-learning-work-in-radiology (accessed August 2022).

20. Kim, S.Y., Choi, Y., Kim, E.K. et al. (2021). Deep learning-based computer-aided diagnosis in screening breast ultrasound to reduce false-positive diagnoses. *Science Reports* 11: 395. https://doi.org/10.1038/s41598-020-79880-0.

21. Gatos, I., Tsantis, S., Spiliopoulos, S. et al. (2016). Quantitative imaging and image processing, a new computer aided diagnosis system for evaluation of chronic liver disease with ultrasound shear wave elastography imaging. *Medical Physics* 43: 1428–1436. https://doi.org/10.1118/1.4942383.

22. Gao, L., Liu, R., Jiang, Y. et al. (2018). Computer-aided system for diagnosing thyroid nodules on ultrasound: a comparison with radiologist-based clinical assessments. *Head & Neck* 40 (4): 778–783.

23. D'Onofrio, M., Beleù, A., Gaitini, D. et al. (2019). Abdominal applications of ultrasound fusion imaging technique: liver, kidney, and pancreas. *Insights into Imaging* 10: 6. https://doi.org/10.1186/s13244-019-0692-z.

24. Scholten, H.J., Pourtaherian, A., Mihajlovic, N. et al. (2017). Improving needle tip identification during ultrasound-guided procedures in anaesthetic practice. *Anaesthesia* 72: 889–904. https://doi.org/10.1111/anae.13921.

25. Xia, W., Noimark, S., Ourselin, S. et al. (2017). Ultrasonic needle tracking with a fibre-optic ultrasound transmitter for guidance of minimally invasive fetal surgery. *Medical Image Computing and Computer-Assisted Intervention – MICCAI* 10434: 637–645. https://doi.org/10.1007/978-3-319-66185-8_72.

CHAPTER 13

Ultrafast Ultrasound

Most of the advanced techniques that have been introduced on ultrasound scanners in recent years are only possible because of advances in computation. In particular

- Speed of the main processors (CPU).
- Memory size and speed of access.
- Graphics processing units (GPUs) capable of massive parallel processing of image and signal data.

Among other things, these developments in computation have been driven by the video games industry. It is now possible to acquire, digitise, and store sequences of complete echo data arriving at all the individual elements of the ultrasound transducer in real time. By complete echo data, we mean the raw RF data of phase and amplitude at ultrasound megahertz frequencies (see Chapter 6). This implies digitising and storing the echo signal at a sample rate of ~40 MHz which quickly generates giga-bytes of data.

This data can then be further processed either in real-time, or offline, once the patient has left the examination room. These advances have led to what has been called **ultrafast ultrasound (UFUS)** [1].

In ultrafast ultrasound, all of the processing, including beamforming, is performed by software. It is therefore not limited by having a restricted number of hardware processing channels.

SYNTHETIC APERTURE IMAGING (SA)

The conventional pulse-echo technique with beam steering and focusing that we have discussed in previous chapters uses **delay-and-sum beamforming**. Pulses from individual elements are delayed so that their phase across the elements used adds up, forming a beam in the direction and with focus as required. In reception, the echoes are similarly delayed and their signals summed to give the received signal. This process is applied to individual pulse-echo sequences to build up the image.

High speed processing enables a novel and conceptually very different way of image formation, called **synthetic aperture imaging**, to be performed that brings lots of additional advantages.

Synthetic aperture imaging is a technique in which an ideal ultrasound source is synthesised by firing the elements of the transducer so as to recreate the beam that would have come from the ideal source. In principle, all of the elements of the transducer are used for each transmission-receive sequence rather than a small group of elements as used in conventional beam forming. In practice, algorithms that use a selection of elements, a **sparse array**, can be used to increase processing speed and computational load whilst maintaining the resolution.

The **ideal source** may be a point source situated some distance behind the ultrasound probe face as shown in Figure 13.1a. The 'point source' can be in any position to allow wavefronts to be sent out in different directions. Another arrangement is to use a source that is effectively at an infinite distance behind the array producing plane waves at the array.

Plane wave beam forming is a common implementation of synthetic aperture imaging in ultrasound and is the form we will explain here. Diverging beam SA imaging is used by phased arrays.

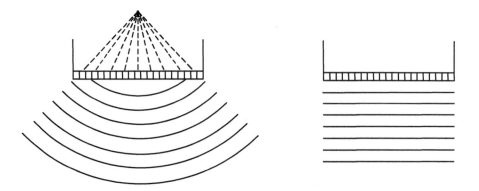

FIGURE 13.1 Examples of an array transmitting wavefronts that would have come from ideal sources. (a) A point source situated behind the array and (b) plane waves from a source situated an infinite distance behind the array.

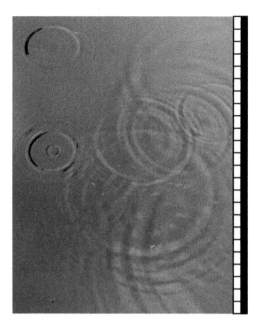

FIGURE 13.2   Raindrops falling in a puddle. Note how the curved wavefronts from each drop position arrive at the edge of the 'pond' (array) in sequence along the edge.

In order to visualise how this technique works, it is useful to watch raindrops falling in puddles or onto a rectangular tray filled with water (Figure 13.2) [2]. Each raindrop creates a wave that moves out from the point where it fell and in time there will be a drop on every point in the puddle. The waves move outwards until they reach the edge, in the same way as echoes returning from the focus of a focused ultrasound beam reach the aperture of an ultrasound transducer. At the edge, there will be a time delay across the edge as each part of the expanding wave meets the edge. The wave from each raindrop point will have a different set of delays in its arrival across the edge that corresponds to the point where it originated. Keep this picture in mind.

PLANE-WAVE BEAMFORMING

A single plane wave pulse is transmitted by firing all the elements across an array (Figure 13.3).

On receive, the phase and amplitude of the echoes arriving back at the transducer are stored for each element in the array, for the entire depth of field of the scan (Figure 13.4). For a single field point P_1, we can take the stored data for each element and retrospectively shift its relative timing so it corresponds to the time delays of the echo from that field point. If we do this for all the elements across the array, we recreate the pulse as it would have arrived back from the point P_1. Then, since data for the whole image field is stored, this can be done for every point in the image and the

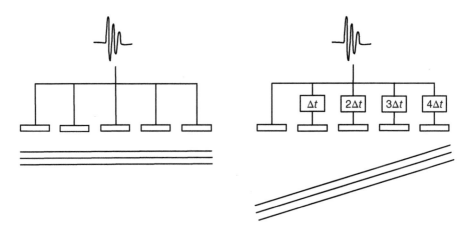

FIGURE 13.3   Transmission of a forward and steered plane wave using the whole array.

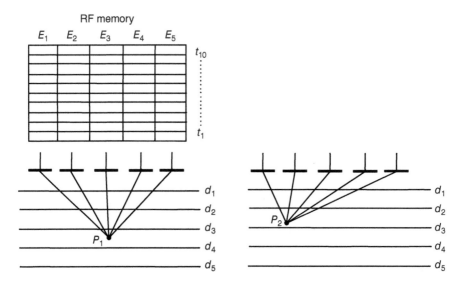

FIGURE 13.4   The RF signal for every element is stored for the whole field of view for each pulse echo sequence. By applying time delays to the stored data, the wavefronts arriving from any field point can be retrospectively reconstructed, e.g. P_1 and P_2.

whole image receive focused. This is just like watching and reconstructing the wavefronts from the raindrops from every point in the puddle.

For example, consider a 5×10 pixel image. The RF data memory stores the phase and amplitude of echoes from each element $E_1, \ldots, E_5$ in the array, arriving at times $t_1, \ldots, t_{10}$.

The data in the memories $E_1, \ldots, E_5$ are combined with focusing delays to focus on a field point P_1 from depth d_4 in the body to give a received focus signal from P_1.

The same is done for, e.g. field point P_2 at depth d_3, and similarly for every other field point corresponding to every point in the image.

If this is done for every scanline (element line) and every depth, then every pixel in the image will be focused on receive.

By this means, a complete receive-focused field of view is obtained from a single transmit pulse.

The computational processing is very fast and for a 20 cm deep field of view and a 128 element array, a **frame rate** of 3.8 kHz can be obtained. This can be compared to a frame rate of 30 Hz for conventional processing.

By combining the stored data in this way, we can produce an image with receive focus for every pixel. However, because the transmission pulse was a plane wave and not focused, image contrast and lateral resolution will be poor. This may be overcome by transmitting several plane wave pulses at different angles from the transducer array on successive transmissions, e.g. $T_1, \ldots, T_3$, and receiving echoes from the same three steered angles (Figure 13.5) [3].

Having received and stored the RF echo data for each element, from each of the transmission pulses, the echo data from each pulse can be delayed and summed with the other transmitted pulses to recreate a transmission focus for every pixel in the image, in the same way as was done for receive focusing of each image pixel (Figure 13.6). It is as though we had run a film of the raindrops backwards so that a set of waves from just three directions converged on each splash point in the puddle.

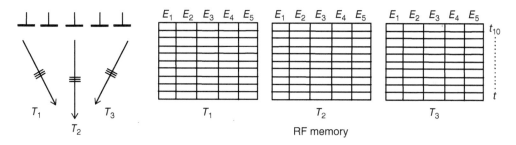

FIGURE 13.5 By transmitting a number of angled beams transmit information is stored for transmit focus reconstruction.

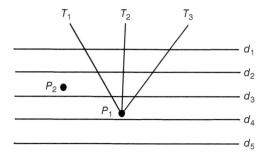

FIGURE 13.6 By applying delays to the stored transmit data for the angled transmitted beams a transmit focus can be reconstructed for each field point in the image.

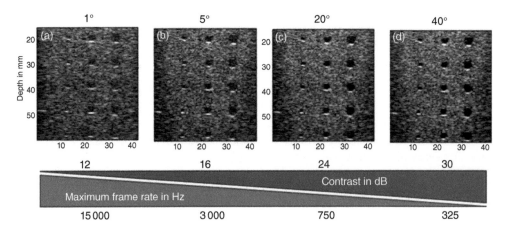

FIGURE 13.7   Relationship between frame rate, number of transmit angles used and image contrast in plane wave transmit focusing. The effect of increasing number of transmit angles may be seen in the images (a–d). *Source:* Bercoff [4]/IntechOpen.

By this means, each pixel in the image can be focused in both transmit and receive and image contrast and resolution are restored.

Having formed the transmit-receive focused data set, the resultant RF data may be processed using the equivalent of envelope detection to obtain the amplitude data for a B-mode image display in the normal way (Chapter 6).

There is a trade-off between frame rate and image quality because several transmit pulses must be transmitted for each frame. However, like-for-like, there is still an increase in frame rate over conventional imaging of 5–10 times faster using plane-wave beamforming. For example, 30 fps conventional becomes 300 fps UFUS (Figure 13.7).

This technique of reconstructing transmit focal points using several angled pulses is called **coherent spatial compounding (SCS)**, as the phase of each pulse is lined up coherently to reconstruct a transmit focus for each point in the field of view.

NOTES

- **Coherent spatial compounding** at RF frequencies enhances resolution by focusing the transmit beam to each field point.
- **Incoherent spatial compounding** of the displayed image amplitude data, as used in conventional B-mode **compound scans**, reduces speckle. This can also be performed in UFUS by collecting several amplitude detected images for different reconstructed transmit angles and compounding them in the normal way.

NOTES

- Lateral resolution is dependent on the maximum angle of the transmit pulses.
- Contrast depends on the number of angles.
- Signal-to-noise ratio (SNR) depends on the number of angles. If the signal-to-noise ratio is made equal to the signal-to-noise ratio for conventional focused ultrasound then with plane-wave beam forming, the signal-to-noise ratio is better across the entire image. In conventional imaging, the SNR is worse outside the focal zone.
- Axial resolution depends on the length of the pulse.
- Focusing on transmit and receive with plane-wave beamforming, together with adaptive beam forming can produce **diffraction limited resolution** across the entire image – i.e. as good as it can get.
- MI and I_{SPTA} are improved as plane wave transmissions are used with no focusing creating zones of high intensity.

SUMMARY

- Transmit plane waves using the full width of the ultrasound array.
- Digitally record the phase and amplitude of the received echoes for every element.
- Reconstruct a receive focus for every image pixel by coherent summation of element signals.
- Reconstruct a transmit focus for every image pixel by coherent summation of several angled transmit pulses.
- Final image quality depends on the number of angled transmit pulses used.
- There is a trade-off between image quality and frame rate.
- Like-for-like in image quality, ultrafast ultrasound is 5–10 times faster than conventional processing, e.g. frame rate 30 becomes 300 frames per second.
- New applications for ultrasound are made possible.

SPEED OF SOUND CORRECTION

Variation in speed of sound leads to general distortion within the image and frank speed of sound artefacts may be seen. Chapter 7 described how poor image quality due to speed of sound aberrations may be improved by modifying the system velocity from 1540 m s^{-1} to one that more closely matches the speed of sound in the tissue being imaged. High-speed processing enables full speed of sound correction to be performed. The principle of these advanced techniques is described. [5]

The average speed of sound in soft tissue is taken to be $1540\,\text{m s}^{-1}$. In reality speed of sound varies around this value by $\pm 5\%$ between different tissue types and any beam path is likely to pass through several tissue types, each with a different speed of sound. This leads to misplacement of the echoes in the image as shown in Figure 13.8.

The principle behind speed of sound aberration correction is to map the variation in speed of sound across the field of view and then use the mapped values to correct the timing of pulses so the echoes for each pixel in the image match their true geometrical location in the image plane.

To understanding how this is done, it is helpful to consider a simple model of a uniform tissue with a speed of sound $C_0 = 1540\,\text{m s}^{-1}$ containing a mass with a different speed of sound C_1 within it (Figure 13.9). If we look at two identical path lengths to a field point P, AP, and BP, in which one crosses the mass and the other does not, there will be a difference in time τ of arrival of echoes along the different paths due to differences in speed of sound along each path.

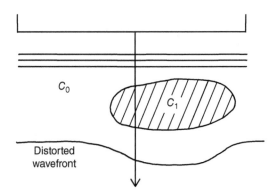

FIGURE 13.8   Distortion of the wavefront caused by a tissue with a slower speed of sound C_1 than the surrounding tissue C_0.

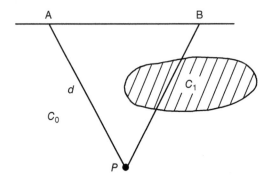

FIGURE 13.9   Simple tissue model showing two identical length paths for the ultrasound beams AP and AB with an inclusion with speed of sound C_1 in a tissue with speed of sound C_0.

For field point P, the 'go and return' time of the echo to A is $t(\text{AP}) = \dfrac{1540}{2d}$
and to B is $t(\text{BP}) = \dfrac{\left(1540 \text{ and } C_1\right)}{2d}$

Because of the tissue mass with speed of sound $C_1 \neq C_0$, there will be a time difference $\tau = \text{BP} - \text{AP}$ in 'go and return' times between the two paths.

Where

$C_1 > C_0$ τ will be negative – echo BP arrives sooner than AP
$C_1 < C_0$ τ will be positive – echo BP arrives later than AP

In the case of real tissue there may be many changes in speed of sound throughout the pulse path and τ will depend on all the changes along the paths AP or BP. The time difference τ is the result of all the changes of speed of sound along those beams. This will be seen as a phase difference between the two signals as shown in Figure 13.10.

By gathering data from many pairs of beam angles, a data set of differences in time of flight τ along the different paths is obtained from which a map of speed of sound can be reconstructed using CT algorithms. This is similar to the way as X-ray CT uses lines of absorption data to produce an X-ray attenuation image. This procedure has been called **computed ultrasound tomography in echo mode (CUTE)**. [6]

The resulting speed of sound map (Figure 13.11) may then be used in one of two ways:

- Diagnostically – the image can be displayed as a colour map to show speed of sound. This gives the sonographer new information that may be diagnostically useful. For example, in breast scanning, lesions may be hard to identify on B-mode but have a high a speed of sound compared with surrounding tissue, and so show up in speed of sound mapping. Liver disease may be seen as a change in speed of sound.
- The speed of sound map may be used to correct the time delays used in beam forming for normal imaging. Using the plane-wave beam forming technique, when focusing for each pixel is performed, the echo data from each element is delayed or advanced using the speed of sound map to correct for the exact time of flight for that echo from that field point. This technique is called **local speed of sound adaptive beamforming**. [7]

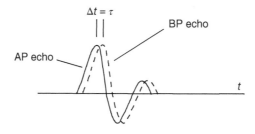

FIGURE 13.10   Phase difference between two echo paths from a field point where $C_1 < C_0$.

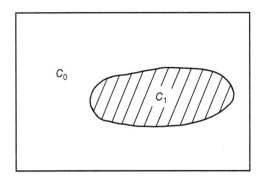

FIGURE 13.11   Speed of sound map constructed from time differences for different field point paths.

Figure 13.12 shows an example of speed of sound correction using CUTE. The images may be compared with Figure 7.9 which shows the effect of speed of sound aberration from the same phantom with and without compound scanning. Note the definition in the image of the wires in each case. Figure 13.13 shows a clinical example of speed of sound correction.

Benefits

Speed of sound correction gives:

- A big improvement in lateral resolution as the reconstructed beams are not distorted by different, unknown, time-of-flight path lengths within the beam-width or within any compound scanning.
- Accurate geometrical placement of echoes in the B-mode image, as speed of sound aberrations have been corrected.
- Improved image quality in both 2-D and 3-D imaging.
- More accurate clinical measurements of dimensions on the image.
- Better matching of ultrasound images to other images in fusion studies.
- Ability to resolve detail at depth in the larger patient.

ULTRAFAST DOPPLER

Ultrafast Colour Doppler (UFCD)

Using the ultrafast technique of transmitting plane waves from multiple angles to create an image in which every pixel is focused also enables Doppler information to be gathered at high speed.[9]

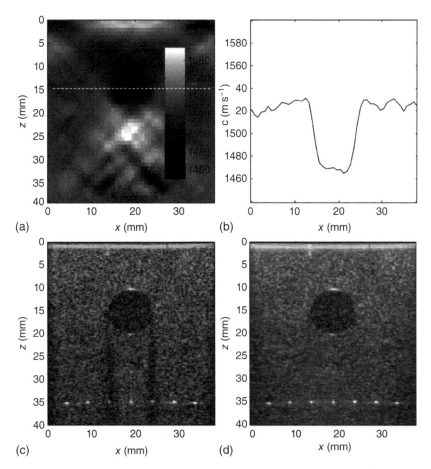

FIGURE 13.12 Using a phantom with an inclusion having a different speed of sound (SoS) (a) shows the SoS map using CUTE, (b) shows the measured SoS across the dashed line, (c) B-mode image from a single forward angle corrected for SoS and (d) a compound scan corrected for SoS. *Source:* (a), (c) and (d) Jaeger et al. [8]/with permission of Elsevier.

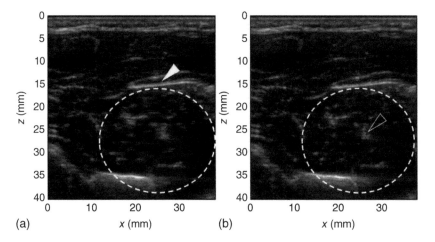

FIGURE 13.13 Speed of sound correction in the forearm (a) B-mode image showing blurring of targets below the fatty layer indicated by the arrow and (b) the same image after correction using CUTE showing uniform lateral resolution throughout the image. *Source:* Jaeger et al. [8]/with permission of Elsevier.

One problem with conventional CDU is that each image line requires ~10 transmit pulses to sample the Doppler information along a line. The **ensemble length** is 10 pulses. This requirement for each colour line reduces the frame rate for CDU to around 10–20 Hz, often restricted to a limited region of the image to improve resolution and frame rate.[10]

To form a CDU image using plane-wave beam forming, plane wave ultrasound pulses are transmitted over a small number of angles at the maximum PRF for the depth of interest (Figure 13.14). In order to get Doppler velocity information that sequence is repeated 10 times. The Doppler image is therefore acquired much faster (e.g. 10–20 times) than for conventional CDU.

As for ultrafast B-mode, RF data is collected for every pulse-echo sequence and stored, e.g. for 5 angles. Sequence T_1 to T_s is then repeated 10 times. This gives an ensemble length of 50 pulses to each field point. All 5N datasets are stored and CDU detection on the dataset, for example, using autocorrelation, is performed to detect the mean velocity in every pixel in the whole image. The sign ($\pm$) of the Doppler frequency, i.e. whether it is higher or lower than the transmit frequency, determines the direction for colour coding as with normal CDU (e.g. red-blue).

In addition to the speed of acquisition there are other advantages:

- Velocity resolution is improved to pixel level.
- Every pixel in the image has colour information, not just a localised colour box.
- There is very good time resolution and all the colour information across the image is derived simultaneously. This removes the time collection artefact seen in large CDU box acquisition (see Chapter 10).
- The ensemble length is effectively much higher than for conventional CDU (e.g. 5×10 pulses = 50 pulses). This gives a less noisy estimate of the Doppler frequency.
- There is increased depth penetration because plane-wave beam forming has increased sensitivity giving a higher SNR.
- There is high sensitivity to low/small flows.
- By selecting various firing sequences for transmit we can optimise the image for high or low flows, or for fast changing flows.

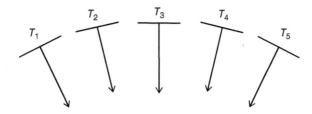

FIGURE 13.14   Plane waves are transmitted over a number of angles to obtain Doppler information.

Microvascular Imaging

Conventional CDU is not able to visualise vessels <1 mm diameter or with very low velocity flows. The problem with imaging such vessels with conventional techniques is that the Doppler shifts from these vessels covers the same frequencies as the motion of the tissue surrounding them. They are therefore excluded by the high pass clutter filter CDU uses to avoid tissue motion artefacts.

In order to image these microvessels, a more sophisticated clutter suppression algorithm must be used. Instead of filtering frequencies in the time domain, it uses relative changes in the speckle pattern of the solid tissue to the speckle pattern of moving blood to discriminate between the two [11]. As the tissue with blood vessels embedded moves, the speckle pattern from a particular point in the tissue moves with the tissue. The speckle pattern from blood in a vessel will also move position but in addition will have an additional relative motion to the moving tissue. The scatterers in the flowing blood have become spatially reorganised to give a new speckle pattern. By applying a spatial filter that can discriminate between speckle that stays the same whilst it moves and speckle that changes as well as moving, the blood motion can be pulled out from the general motion of the tissue and displayed in colour on the image. To detect these small differences requires the tissue/flows to be sampled many times (colour frame rates of 50–60fps) and it is the advent of UFUS that has made this possible. Further processing to remove image degradation due to tissue motion may be applied, such as combining sequences of pulses whose phase and amplitude are changed. This becomes important for looking at very low flows where longer acquisition times are needed. The end result is that low flows in very small vessels can be visualised. Examples are shown in Figures 13.15 and 13.16.

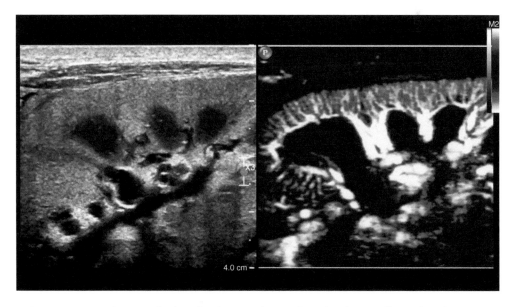

FIGURE 13.15 Microvascular imaging in a renal transplant showing small vessels right out to the external border of the cortex. *Source:* Illustration by courtesy of Koninklijke Philips N.V.

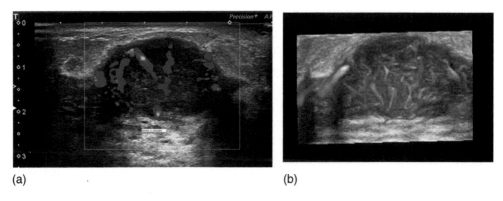

(a)   (b)

FIGURE 13.16   Vascularity in a ductal carcinoma of the breast. (a) CDU image and (b) microvascular 3D image. *Source:* Zhang et al. [12]/Dove Press Ltd./CC BY NC 3.0.

NOTE

Microvessels may be imaged by this technique without the need for contrast agents although contrast agents can improve visibility and SNR.

Pulsed Wave Doppler (PW)

With UFCD, a PRF as high as that used for conventional pulse Doppler can be achieved for the whole image. This means that a PW Doppler sonogram can be produced for any pixel in the UFCD image simultaneously with the CDU display. There is no degradation of the CDU image or sonogram as is the case with such time sharing with conventional CDU. Furthermore, a sonogram may be shown for multiple points within a blood vessel simultaneously (Figure 13.17). If the UF Doppler data set is obtained and stored across a complete cardiac cycle, then the CDU map may be retrospectively shown for any point in the cycle with multiple PW sonograms chosen and shown with the cycle point marked on them.

NOTES

- Using vector flow imaging, the Doppler angle for the sonogram is automatically determined by the scanner.
- PW Doppler shows much reduced intrinsic spectral broadening than conventional PW [14].

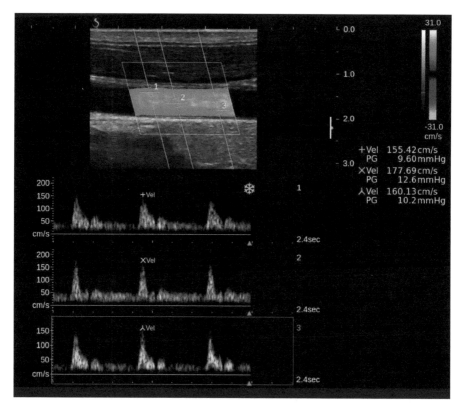

FIGURE 13.17 Illustrating the principle of showing multiple sonograms simultaneously. *Source:* Bercoff et al. [13]/SuperSonic Imagine S.A.

VECTOR FLOW IMAGING (VFI)

In order to completely characterise the flow in the heart or in blood vessels it would be useful to show the direction and speed of blood flow at every point in the vessel. This is what the display of **flow vectors** does. It can yield information on areas of turbulence, enable **volume** flow to be measured and enable **pressure drops** across a stenosis to be estimated noninvasively [15, 16].

The Flow Vector

To fully specify the motion of a particle of blood we need to know both its **speed** and **direction** of travel. These two together form the **flow vector v** for that particle. (The bold 'v' indicates that it is a vector rather than a simple velocity or speed.)

We can relate the flow vector of the particle to the axes of the ultrasound image where X is the transverse axis across the image and Z is the depth in the image away from the transducer as shown in Figure 13.18.

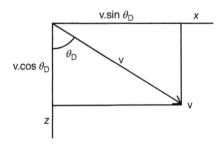

FIGURE 13.18   The flow vector **v** at Doppler angle θ_D showing the x and z components of velocity.

θ_D is the **angle of the flow** from the z-axis and is the familiar Doppler angle. The **length** of the vector **v** shows the **speed** or **magnitude of velocity**. The component of speed v_z in the Z direction is given by $v.\cos \theta_D$, which appears in the normal Doppler equation, and the component of speed in the X direction v_x is given by $v.\sin \theta_D$.

In order to determine the vector of a blood particle in the image we need to find both its velocity in the Z direction and in the X direction. Standard Doppler measurement finds the velocity in the Z direction as given by the Doppler equation

$$V_z = \frac{cf_D}{2f_T}\cos\theta_D$$

As shown in Chapter 10, CDU estimates the velocity in the Z direction by comparing the RF signal from several successive pulses to see how the echo pattern has moved between them using the method of autocorrelation. What conventional CDU shows is a map of v_z and $\cos \theta_D$ remains unknown. If the ultrasound beam was aligned along the direction of flow, then θ_D would be zero and $\cos \theta_D = 1$ and we would have a colour flow map that showed real velocity of flow along the vessel.

Directional Beam Forming

Using plane-wave beamforming, we can form a beam in any direction, including parallel to the probe face (Figure 13.19). This technique is known as **directional beamforming** [17, 18]. By focusing along a line in the image we create such a beam. That is, we use the UFUS focused data to look at points along a line in the image. So, for example, using the stored RF data from the plane wave transmissions, we can beamform along the line AB by forming focuses along the line AB, as shown. Similarly, we could beamform along line CD, parallel to the probe face. If we beam form along the line of a blood vessel and use cross-correlation to compare the echo patterns along that line, we detect the movement along the vessel with no $\cos \theta_D$ term, in other words, we detect v_{flow} along the vessel.

In order to use this procedure, we still need to know the line of the vessel in the image plane. [19] The way to find this is to create a series of beamformed lines at a

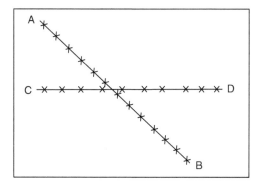

FIGURE 13.19   Illustration of a set of focuses set to form two beams as shown.

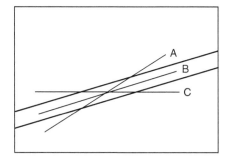

FIGURE 13.20   Using a series of directional beamformed lines to determine the angle of a vessel.

set of angles, e.g. 2° apart, and cross-correlate along each of the lines (Figure 13.20). The line that yields the highest correlation will be the line that is in line with the flow. Using the highest correlating values across the image plane yields a flow map in which the flow velocity and angle of flow is known for all points. By this means, we can form a **vector flow map** as shown in Figure 13.21.

The problem with directional beamforming in this way is that it is computationally very expensive. A whole set of lines need to be beamformed and tested for maximum correlation each time each image point is refreshed. An alternative technique allows the angle of flow θ_D to be determined directly and quickly for each field point, but gives a slightly lower accuracy for the final value of the flow vector. The technique is called **transverse oscillation** and it allows the lateral v_x component of the flow to be determined.

Transverse Oscillation (TO)

As noted above, the normal CDU map shows the value of v_z along the axis of the ultrasound beam, as cross-correlation between successive pulses shows movement along the beam axis.

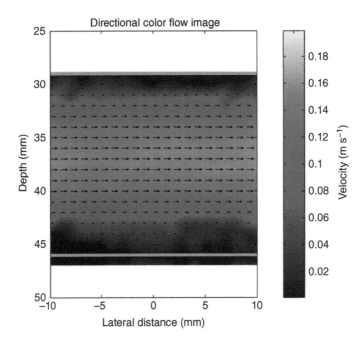

FIGURE 13.21   A vector flow map of a vessel at 90° to the surface of the probe using directional beamforming. Arrows show the vectors of velocity. *Source:* Jensen and Oddershede [19]/with permission of IEEE.

The estimation of v_x, the transverse velocity component of blood, is a more difficult problem. One way to find the transverse velocity would be to look for transverse movement in the speckle pattern across the image from one frame to the next. This depends on the frame rate and is not particularly accurate.

A better way is to create a 'pulse' in which the wavefronts are aligned along the z-axis and travel out along the X axis, in a similar manner to the way normal pulse wavefronts are aligned along the X axis and travel down the Z axis. That is, we create a **transverse oscillation** of normal longitudinal pressure waves as shown in Figure 13.22.

This then enables the transverse velocity v_x to be estimated as changes in the phase of transverse oscillations, in the same way as conventional CDU estimates v_z from phase changes along the beam axis (see Chapter 9).

The principal of how transverse oscillations are created is as follows (Figure 13.23).[20] If two beams of ultrasound are transmitted at the same time so they crossover at a point in the image plane they will interfere with each other to create wavefronts aligned along the z-axis.

The wavelength of the transverse oscillations in the x-direction is given by

$$\lambda_x = \frac{2\lambda_T z}{d}$$

The x and z components of velocity can then be detected by beamforming – equivalent to sending out pulses – along three directions. The first beam is sent out

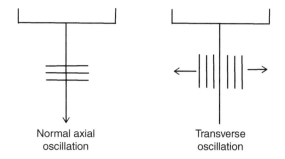

Normal axial
oscillation

Transverse
oscillation

FIGURE 13.22   Diagram showing the difference between axial and transverse oscillation.

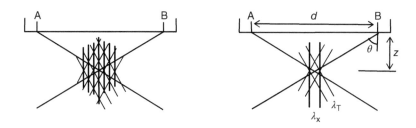

FIGURE 13.23   The creation of transverse waves by using interference from two crossing beams A and B. (a) shows how the normal wavefronts perpendicular to each beam interfere to produce wavefronts in the transverse (x) direction. (b) shows the resultant transverse wavelength λ_x compared to the transmitted wavelength λ_T.

along the z axis and obtains a Doppler sample for v_z in the conventional way. The other two beams are sent out together so as to cross over and create a transverse oscillation. The detected phase – equivalent to the echo back – will give a Doppler sample of the v_x component.

The Doppler signals for f_z and f_x are estimated by comparing the phase shift between successive sets of three beamformed pulses using autocorrelation between sample pairs. Then

$$f_z = \frac{2 v_z f_T \cos\theta}{c} \quad \text{and} \quad f_x = \frac{v_x}{\lambda_x}$$

From which the velocity vector for each pixel in the image can be calculated using

$$\theta_D = \arctan\left(\frac{v_x}{v_z}\right) \quad \text{and} \quad v = \sqrt{v_z^2 + v_x^2}$$

The two interfering beams are apodised to give a **Gaussian beam shape** (see Chapter 3) so that side lobes, that could give erroneous information, are reduced and the beams have a uniform shape at all depths of interest.

Using plane-wave beam forming, the whole field of view can be scanned for z-axis and x-axis vector information at real-time rates with transverse oscillations being formed for every field point in the image from the stored RF data.

The resolution of transverse oscillation detected vectors is somewhat less than that using directional beamformed scan lines. The two can be combined by using TO to find the vector then using directional beamforming to give the displayed vectors. [21]

Display of Flow Vectors

Determining the magnitude and direction of flow in every part of the image provides an enormous amount of data and there is the question of how to display this data in a meaningful way that can easily be appreciated by the sonographer. Two displays that have been used are a vector display and a colour map display.

Vector Display

Vectors are shown as small arrows superimposed on the blood flow such that the length and direction of the arrow shows the speed and direction of flow (Figure 13.24).

Colour Display

Flow direction and speed may be shown using a colour mapping that has direction shown by colour hue and velocity by colour saturation (Figure 13.25).

Using 2-D transducer arrays, vector flow mapping can be extended to 3-D vector flow maps (Figure 13.26). Data may be collected across the cardiac cycle enabling flow velocities to be examined at each stage in the cycle to give 4-D information. [25]

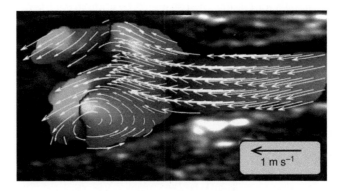

FIGURE 13.24   A flow vector display using arrows to show flow direction superimposed on a normal CDU map. *Source:* Madiena et al. [22]/with permission of IEEE.

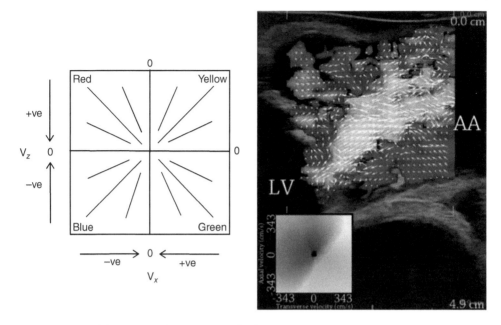

FIGURE 13.25 (a) Diagram of display scale for flow vectors using colour and saturation to show direction and magnitude of flow. (b) example of jet flow from an aortic valve AV stenosis into the ascending aorta AA. *Source:* Hansen et al. [23]/with permission of Elsevier.

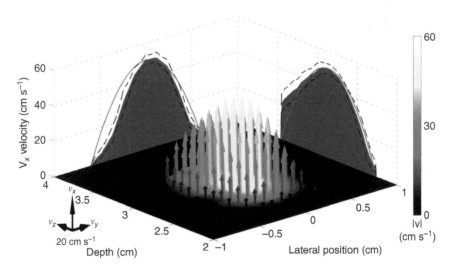

FIGURE 13.26 Velocity vectors at one point in a time sequence from a 3D scan of an 8mm vessel at 90° to the probe face. *Source:* Holbek et al. [24]/with permission of IEEE.

> **NOTES**
>
> - Knowing the velocity vector within the vessel, the scanner can automatically set the correct angle for pulse wave (PW) Doppler waveforms.
> - Knowing the velocity distribution across the vessel lumen, the volume flow (mL min^{-1}) can be calculated.
> - Knowing the velocity distribution at all points along a vessel or across a valve enables pressure gradients within the vessel to be estimated noninvasively. For example the pressure drop caused by a tight stenosis. With velocity vector data, this may be performed using the Navier-Stokes equation to give better estimates of pressure drop than is possible with conventional imaging (see Chapter 12) [26].

REFERENCES

1. Tanter, M. and Fink, M. (2014). Ultrafast imaging in biomedical ultrasound. *IEEE Transactions on Ultrasonics, Ferroelectrics, and Frequency Control* 61: 102–119.
2. The American Registry for Diagnostic Medical Sonography (2022). https://www.ardms.org/dr-kremkau/ (accessed August 2022).
3. Montaldo, M., Tanter, M., Bercof, F.J. et al. (2009). Coherent plane-wave compounding for very high frame rate ultrasonography and transient elastography. *IEEE Transactions on Ultrasonics, Ferroelectrics, and Frequency Control* 56: 489–506.
4. Bercoff, J. (2011). Ultrafast ultrasound imaging. In: *Ultrasound Imaging - Medical Applications* (ed. I.V. Minin and O.V. Minin) [Internet]. London: IntechOpen https://doi.org/10.5772/19729.
5. Jaeger, M., Robinson, E., Akarçay, H.G. et al. (2015). Full correction for spatially distributed speed-of-sound in echo ultrasound based on measuring aberration delays via transmit beam steering. *Physics in Medicine and Biology* 60: 4497–4515.
6. Stähli, P., Kuriakose, M., Frenz, M. et al. (2020). Improved forward model for quantitative pulse-echo speed-of-sound imaging. *Ultrasonics* 108: 106168. https://doi.org/10.1016/j.ultras.2020.106168.
7. Rau, R., Schweizer, D., Vishnevskiy, V. et al. (2019). Ultrasound aberration correction based on local speed-of-sound map estimation. *IEEE International Ultrasonics Symposium* 2019: 2003–2006. https://doi.org/10.1109/ULTSYM.2019.8926297.
8. Jaeger, M., Held, G., Peeters, S. et al. (2015). Computed ultrasound tomography in echo mode for imaging speed of sound using pulse-echo sonography: proof of principle. *Ultrasound in Medicine and Biology* 41: 235–250. https://doi.org/10.1016/j.ultrasmedbio.2014.05.019.
9. Bercoff, J., Montaldo, G., Loupas, T. et al. (2011). Ultrafast compound doppler imaging: providing full blood flow characterization. *IEEE Transactions on Ultrasonics, Ferroelectrics, and Frequency Control* 58: 134–147.
10. Jensen, J.A., Nikolov, S.I., Hansen, K.L. et al. (2019). History and latest advances in flow estimation technology: from 1-D in 2-D to 3-D in 4-D. *IEEE International Ultrasonics Symposium* 2019: 1041–1050. https://doi.org/10.1109/ULTSYM.2019.8926210.
11. Demené, D., Deffieux, T., Pernot, M. et al. (2015). Spatiotemporal clutter filtering of ultrafast ultrasound data highly increases doppler and fUltrasound sensitivity. *IEEE Transactions on Medical Imaging* 34: 2271–2285.

12. Zhang, X., Zhang, L., Li, N. et al. (2019). Vascular index measured by smart 3-D superb microvascular imaging can help to differentiate malignant and benign breast lesion. *Cancer Management and Research* 11: 5481–5487.

13. Bercoff. J., Chamak-Bercoff. J., Fraschini. C. et al. (2011). UltraFast™ doppler. Supersonic Imaging White Paper.

14. Malonea, A., Cournane, S., Fagan, A. et al. (2019). Investigation of intrinsic spectral broadening on the Aixplorer ultrafast ultrasound system. *Physica Medica* 67: 210.

15. Jensen, J.A., Nikolov, S.I., Yu, A.C.H. et al. (2016). Ultrasound vector flow imaging: part I: sequential systems. *IEEE Transactions on Ultrasonics, Ferroelectrics, and Frequency Control* 63: 1704–1721. https://doi.org/10.1109/TUFFC.2016.2600763.

16. Jensen, J.A., Nikolov, S.I., Yu, A.C.H. et al. (2016). Ultrasound vector flow imaging: part II: parallel systems. *IEEE Transactions on Ultrasonics, Ferroelectrics, and Frequency Control* 63: 1722–1732. https://doi.org/10.1109/TUFFC.2016.2598180.

17. Jensen, J.A. and Nikolov, S. (2002). Transverse flow imaging using synthetic aperture directional beamforming. *IEEE Ultrasonics Symposium Proceedings* 2002: 1488–1492. https://doi.org/10.1109/ULTSYM.2002.1192586.

18. Jensen, J.A. and Nikolov, S.I. (2004). Directional synthetic aperture flow imaging. *IEEE Transactions on Ultrasonics, Ferroelectrics, and Frequency Control* 51: 1107–1118.

19. Jensen, J.A. and Oddershede, N. (2006). Estimation of velocity vectors in synthetic aperture ultrasound imaging. *IEEE Transactions on Medical Imaging* 25: 1637–1644. https://doi.org/10.1109/TMI.2006.883087.

20. Jensen, J.A. and Munk, P. (1988). A new method for estimation of velocity vectors. *IEEE Transactions on Ultrasonics, Ferroelectrics, and Frequency Control* 45: 837–851.

21. Jensen, J., Hoyos, V., Stuart, C.A. et al. (2017). Fast plane wave 2-D vector flow imaging using transverse oscillation and directional beamforming. *IEEE Transactions on Ultrasonics, Ferroelectrics, and Frequency Control* 64: 1050–1062. https://doi.org/10.1109/TUFFC.2017.2693403.

22. Madiena, C., Faurie, J., Poree, J. et al. (2018). Color and vector flow imaging in parallel ultrasound with sub-nyquist sampling. *IEEE Transactions on Ultrasonics, Ferroelectrics, and Frequency Control* 65: 795–802. https://doi.org/10.1109/TUFFC.2018.2817885.

23. Hansen, K.L., Moller-Sorensen, H., Kjaergaard, J. et al. (2016). Intra-operative vector flow imaging using ultrasound of the ascending aorta among 40 patients with normal, stenotic and replaced aortic valves. *Ultrasound in Medicine and Biology* 42: 2414–2422.

24. Holbek, S., Christiansen, T.L., Stuart, M.B. et al. (2016). 3-d vector flow estimation with row–column-addressed arrays. *IEEE Transactions on Ultrasonics, Ferroelectrics, and Frequency Control* 63: 1799–1814. https://doi.org/10.1109/TUFFC.2016.2582536.

25. Correia, M., Provost, J., Tanter, M. et al. (2016). 4D ultrafast ultrasound flow imaging: in vivo quantification of arterial volumetric flow rate in a single heartbeatPhys. *Medical Biology* 61: L48–L61.

26. Olesen, J.B., Villagómez, H.C.A., Møller, N.D. et al. (2018). Non-invasive estimation of pressure changes using 2-D vector velocity ultrasound: an experimental study with in-vivo examples. *IEEE Transactions on Ultrasonics, Ferroelectrics, and Frequency Control* 65: 709–719. https://doi.org/10.1109/TUFFC.2018.2808328.

CHAPTER 14

Elastography

The tissues in the body that are loosely termed 'soft tissues' vary in their stiffness/rigidity or their elasticity. And the stiffness of the tissue may change when it becomes diseased, for example, a malignant tumour is often stiffer than surrounding tissue, or in response to physiological changes such as oedema. Measuring tissue **stiffness** or **elasticity** therefore provides another 'window' into the body to form a diagnosis.

Elasticity is a measure of how much the tissue moves in response to an applied force. **Rigidity** or **stiffness** is a measure of how much the tissue resists moving in response to an applied force,

$$\text{i.e. elasticity} = \frac{1}{\text{rigidity}}$$

The basic principle behind elastography is that we apply a deforming force to create a distortion in the tissue and from the way it responds we can infer its mechanical properties and display the results, for example, as a colour mapping on a B-mode image.

For an everyday example, to see how this principle works, imagine pressing down on the surface of a trifle in a glass bowl with a spoon. The custard on the top has very little rigidity and moves easily. The jelly requires more force to squeeze it and the cake in the jelly is relatively rigid and does not distort much at all.

Elastography with ultrasound can be thought of as a more sophisticated extension of manual palpation, which is regularly used by physicians to examine patients.

BACKGROUND THEORY

Within a time frame of <1 second, soft tissue may be considered incompressible. If a force is applied in one direction, the dimension of the tissue on that axis will decrease but the volume remains the same as there will be a corresponding spreading out in a plane lateral to the applied force, known as the **Poisson effect** (Figure 14.1). When the force is removed, the tissue rebounds to its original shape. If the force is sustained, then the property of incompressibility may no longer be true. For example, when compressing oedematous tissue, fluid may be driven out of the compressed tissue into the surrounding tissue space and the oedematous tissue then has a reduced volume until it refills and induration of the skin may be noticed.

Assuming soft tissue is incompressible, if we press down on the skin, the tissues will move a certain amount for a particular applied force. The deformation will vary from one tissue to another depending on how stiff it is. The distance along the axis of the applied force will decrease and the tissue will spread out and expand in a direction perpendicular to the applied force as described.

The applied force is called the **stress** σ and the resulting change in length along the axis of the applied force, $\dfrac{\Delta l}{l}$ is called the **strain** (Figure 14.2). The ratio of the two is called **Young's modulus** E.

$$E = \frac{\text{stress}}{\text{strain}} = \frac{\text{applied force}}{\text{resulting movement}}$$

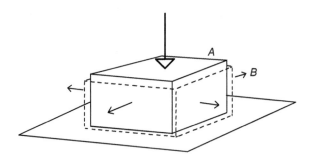

FIGURE 14.1   The motion of tissue from A to B under an applied unidirectional force assuming it is incompressible.

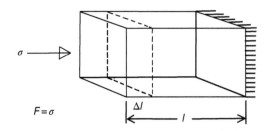

FIGURE 14.2   Definition of Youngs modulus: change in length Δl in response to applied force σ.

E has a high value for a stiff rigid material (small strain) and a low value for a soft deformable material (large strain).

So, by applying a force and using the change in position of the target tissue, measured from the returning ultrasound echoes, we can estimate the stiffness of tissues in the body.

ELASTOGRAPHY

The general term elastography is used to cover a set of closely related ultrasound techniques that give information on tissue stiffness. These fall into two groups: those that estimate the strain resulting from an applied force are called **strain elastography (SE)**, and those that create a sideways moving displacement or **shear wave** inside the tissue, and measure the velocity of the shear wave to estimate tissue stiffness, are called **shear wave Elastography (SWE)**. [1, 2]

METHODS OF APPLYING THE DISTORTING FORCE

In order to measure tissue stiffness with ultrasound, we need to apply a force to distort the tissue. There are three main ways to achieve this:

1. Push or vibrate the surface of the body with the ultrasound probe or tool such as a probe footplate extender.
2. Use distortions created by normal physiological movements, e.g. pulse, respiration, or heartbeat.
3. Use acoustic radiation force from an ultrasound pulse to remotely create a 'push' inside the tissue.

The results obtained may be qualitative – showing relative stiffness, or they may be quantitative – giving actual stiffness as a numerical value.

STRAIN ELASTOGRAPHY (SE)

Uses the change in position from an applied force to estimate strain.

The applied force is provided either by manual compression on the skin, e.g. by pushing down on the transducer probe, or by tissue displacement from the heart or respiration, as may be used when looking at breast lesions.

SE with Manual or Physiological Palpation

If a constantly increasing force is applied such that the strain increases uniformly, speckle tracking across successive frames can detect the movement of the tissue.

Speckle Tracking

The speckle pattern, produced by the interference of microscopic scatterers within the length of the ultrasound pulse, will remain static as long as the tissue and probe remain stationary. If the tissue moves, even by a fraction of a wavelength, the speckle pattern will change as a new interference pattern emerges from a new pattern of scatterers. However, as long as the movement in tissue is small, of the order 1 wavelength, the speckle pattern from a particular group of scatterers will be virtually unaltered but will occur at a slightly different time in the pulse-echo sequence (Figure 14.3). Speckle tracking compares the received RF signal from consecutive pulses to find the speckle pattern from the first pulse in the second and so determine how far the speckle pattern has moved using

$$\Delta d = \frac{1540.\Delta t}{2}$$

where Δt is the time between pulses.

Tracking the speckle in this way can be achieved either by using autocorrelation to find the best RF pattern match between pulses, or with Doppler techniques which look for the phase difference between consecutive pulses.

Speckle tracking may be carried out in the axial direction by comparing pulses along the same beam or in the lateral direction by comparing pulses from adjacent beams.

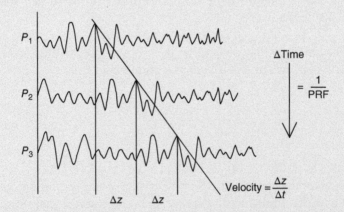

FIGURE 14.3   Speckle tracking across three consecutive pulses P_1-P_3. The change in position is detected by looking for identical patterns in the RF signal, from which the distance or velocity of movement can be estimated.

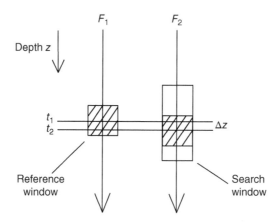

FIGURE 14.4 Speckle tracking to determine strain. The RF signal from a reference window in the first frame F_1 is looked for within a search window in the second frame F_2.

Using the RF echo signal and cross-correlation, a reference window from the first frame is compared across the search window in the second frame to find the best fit (Figure 14.4). For each scan line, the axial distance Δz that the reference window signal has moved is the displacement caused by the applied force. The procedure is performed for all depths along each scan line so as to cover the region of interest. This is repeated using successive pairs of frames with frame averaging, to improve the estimated value of the displacement (Figure 14.5). (**Note**: this is not the same window as the elastogram box seen on the B-mode image)

From the displacement obtained from the **strain-estimating window** or **tracking window,** the strain is then calculated as the gradient of displacement, as is seen in Figure 14.6.

$$\text{strain} = \frac{\Delta d}{d_0} \cdot 100\% \quad \text{i.e. the gradient as a percentage change, where } d_0 \text{ is the original}$$

spacing of the echo pattern.

The size of the window is a compromise between resolution and signal to noise ratio (SNR), and is a user control on some machines.

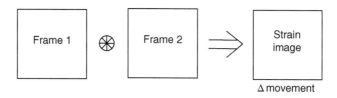

FIGURE 14.5 The principle of comparing two successive images to look at movement in the image frame.

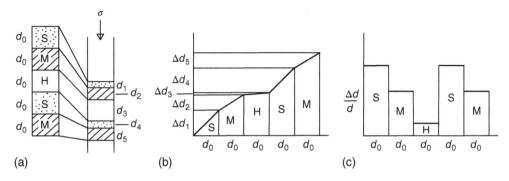

FIGURE 14.6   (a) The change in thickness of a tissue with 5 layers, thickness d_0, but whose stiffness varies (H hard, M medium and S soft) when a stress σ is applied. (b) The change in thickness ($\Delta d_1 = d_0 - d_1$) graphically. (c) The gradient for each layer shows the strain.

On compression by the transducer, there may be tissue displacement that is lateral to the beam axis but still within the scan plane. That is, axial compression causes some lateral expansion that causes displacement noise (see Figure 14.1). The scanner therefore performs cross-correlation in the lateral and axial directions (i.e. across beams Δx as well as along beams Δz) to improve the precision of the axial displacement estimate, but only retains the axial component for the elastogram. This reduces noise arising from lateral tissue movement.

The resulting strain data may be displayed as a colour map or a greyscale map displayed in an **elastogram box** on the B-mode image. The map is known as an **elastogram**. The size of the elastogram box may be chosen by the user. In order to improve the quality of the elastogram, frame averaging is applied.

The use of a footplate extender on the transducer can improve the uniformity of the applied stress and maximise the depth to which it penetrates (Figure 14.7)

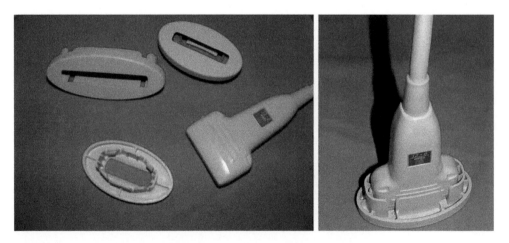

FIGURE 14.7   Probe footplate extender. *Source:* Dietrich et al. [3]/Thieme Medical Publishers, Inc.

> **NOTE**
>
> There is not a standardised colour scheme and colours may vary between scanners, e.g. red-green-blue – may be hard/soft or soft/hard. USER BEWARE!.

USER CONTROLS

> **NOTE**
>
> Not all scanners will have these controls accessible to the user.

Frequency

The ultrasound frequency used for the elastogram may be chosen independently of the B-mode frequency. As for B-mode, a lower frequency will give greater penetration for sensing the strain but poorer resolution. Reducing the frequency also reduces the elastogram frame rate.

Frame Rate

Some scanners have a separate control that adjusts the elastogram frame rate to enable the user to match the compression speed and amplitude.

Tracking Window

The tracking window used for speckle tracking may be adjusted. A larger window gives a better signal to noise ratio but reduces the resolution.

Frame Reject

Lower quality elastogram frames are rejected and colour is not shown in the elastogram box for these frames. This can lead to a 'flashing' effect on the elastogram as frames are rejected or accepted. Lowering the reject threshold reduces flashing but also reduces the quality of the elastogram.

Noise Reject

A filter that removes noisy pixels within each frame by rejecting regions where the echo signal is not strong enough for speckle tracking correlation, for example, in cysts and other hypoechoic areas. Colour is not displayed in rejected pixels.

Line Density

A lower line density results in a higher frame rate but with lower spatial resolution.

Frame Averaging

Allows fames to be blended with previous frames to improve image quality. It reduces flashing.

Dynamic Range

Adjusts the levels at which significant colour changes indicating strain levels are made. With higher values of dynamic range only more extreme values of strain are shown in red or blue with more of the image showing green.

Colour Map

The colour map used to display the elastogram may be changed.

Colour Blend

Alters the transparency of the elastogram box over the B-mode image.

NOTES

- Hand compression motion must be in line with the axis of the ultrasound beams. If there is movement along a line that runs out of the scan plane or twists with respect to a fixed scan plane then cross-correlation between frames will be lost and the elastogram will be noisy or dropout.
- Uneven stress across the footprint of the transducer leads to non-uniformity in the elastogram. Therefore 'heel and toe' movements of the transducer should be avoided.
- Curved array transducers present a problem due to tissue under the centre of the transducer experiencing higher stress than that at the ends of the transducer. This can be mitigated by the use of a probe footplate to increase the surface area of the transducer and give more consistent pressure along its length.
- The best quality data comes from echoes where there is near constant rate of surface displacement, not at the extremes of displacement, as shown in Figure 14.8.
- The strain images should be optimised by the sonographer varying the **compression speed** (0.5/second–2.0/second), the **amplitude** (1–3 mm) of the compression and the persistence or **frame averaging** used. These should

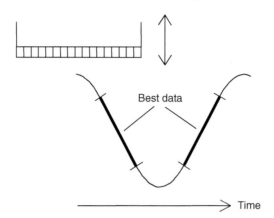

FIGURE 14.8 The best SE data comes from uniform compression of tissue.

be adjusted interactively whilst viewing the image so as to best display the tissue strain contrast between hard and soft tissue. The image **frame rate** may also be altered to achieve best strain contrast.

- Some scanners have a **quality indicator** giving a numerical or visual display of the amount of compression-release being applied to aid the operator in finding the optimal technique. More recently indicators of pressure gradient across the transducer footprint have become available enabling the operator to check they are applying an even pressure.
- Aim for consistency in strain seen across frames.
- The size of the **elastogram box** and the line density may be adjusted to improve frame rate. However, the box should include enough normal tissue so that relative changes to normal, in abnormal tissue, can be observed. The suspected lesion should take up <50% of the elastogram box [3].
- Care should be taken when adjusting the frame reject and noise reject controls as they determine the information that is, or is not, seen in the strain image.
- The strain image shows relative differences in stiffness and the information is qualitative. Comparison should not be made across different manufacturers scanners.
- Strain imaging works best where there are **focal changes** in tissue stiffness within the field of view. It shows relative contrast between tissue stiffness and therefore gives little information where there is a generalised change in tissue stiffness such as is seen in cirrhotic liver, where the lack comparator tissues means relative strains are not seen.
- Imaging at depth depends on the transmission of the palpation into the tissue. The displacement reduces with depth but the strain, or percentage deformation, will be constant with depth. A deeper scan may necessitate a slower rate of probe compression and hence reduce frame rate.

The small tissue movements resulting from cardiac pulsation and respiration may also be tracked to produce strain images, e.g. in breast tissue.

SE ARTEFACTS

Care needs to be taken in interpreting strain images as a number of physical features can affect the elastogram causing artefacts.

- Tissue is non-linear in the relationship between the force applied and the deformation produced leading to artefacts such as stress concentration artefact.
- Elasticity in tissue is often **anisotropic** and probe orientation with respect to the tissue may alter the result.
- The tissues in the body are mechanically discontinuous when interrupted by anatomical features such as fluid collections, organ or tumour boundaries and scars. These may produce artefacts on the strain image.
- As tissue is compressed it becomes more rigid and **pre-compression** or **pre-loading** by applying too much pressure with the probe prior to starting compressions, will diminish the strain contrast and produce a noisy image. The pressure should be only just sufficient to obtain a diagnostic B-mode image. A generous layer of lubricating gel will enable good contact with little initial pressure and good B-mode to be obtained.

The graph in Figure 14.9 shows the effect of pre-compression on strain contrast in breast imaging. Cancerous tumours are stiffer than fat. If the breast tissue

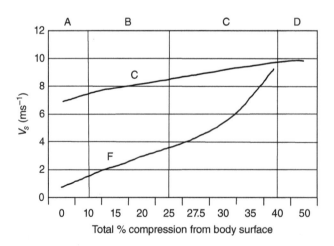

FIGURE 14.9   The effect of pre-compression on shear wave velocity v$_s$ in the breast for fat (F) and cancer (C). In zone A (<10% pre-compression) benign and malignant tissue may be distinguished on strain and shear wave images. In zones B-D the signal degrades. Strain images become too noisy to use and shear wave velocities become indistinguishable. *Source:* Adapted from Barr and Zhang [4].

is compressed by the probe so that the measured thickness of the breast tissue from skin to a fixed point (e.g. rib) is reduced by 40%, the fat shows the same stiffness as the tumour. The effect on the elastogram is shown in Figure 14.10.

As a result of these factors, it is recommended that the following be observed where possible when performing SE.

- Strain imaging is only going to be reliable if there is a good quality B-mode image. Therefore, **ensure you have a good B-mode image**.
- Targets closer to the transducer are imaged more reliably.
- Targets surrounded by near homogenous tissue are imaged more reliably than where there are closely adjacent tissue boundaries.
- There should be no anatomical planes that allow slip movement anterior to or within the region of interest (Figure 14.11).

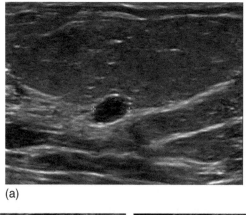

(a)

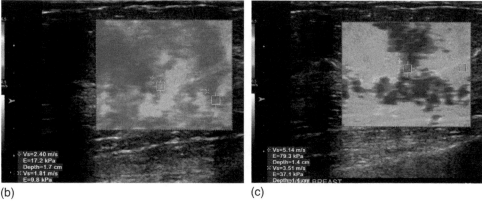

(b)   (c)

FIGURE 14.10 SW Elastogram showing the effect of pre-compression. (a) B-mode of a benign fibroadenoma, (b) its shear wave velocity when done properly is 2.4 ms^{-1} and (c) with heavy pre-compression rises to 5.1 ms^{-1}. Note also, the shear wave velocity of surrounding fat also increases with increased compression. *Source:* (b) and (c) Barr [5]/Korean Society of Ultrasound in Medicine.

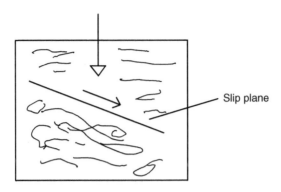

FIGURE 14.11   Diagram of a tissue boundary that may form a slip plane when performing SE.

- Check the B-mode image to make sure there is no lateral movement of the tissue with the compression or breathing, etc.
- There should be no structures nearby that would damp the shear stress, e.g. large veins.
- Aim for a limited number of targets in the elastogram field of view so it covers normal and abnormal tissue (target <50% of elastogram box area). [6]

The following artefacts may be seen on SE. Some of these artefacts can be diagnostically useful or alert the sonographer to technical problems in the scan.

Stress Concentration Artefacts

Soft tissue strain is greater when it is adjacent to hard tissue and the inclusion of a hard tissue in a soft background may cause the surrounding soft tissue to appear softer than it is where **stress concentration** occurs as shown in Figure 14.12.

Edge Enhancement Effects

Slippery boundaries between structures may cause edge enhancement effects to be seen on the elastogram. This occurs where, upon palpation, tissue on one side of a boundary slips relative to the tissue on the other side (Figure 14.11). Or, it may occur when a solid structure is not strongly tethered and moves with respect to surrounding tissue with the compressions (Figure 14.13). As a result, the edge may appear enhanced and there may be slight displacement of strain boundaries relative to the corresponding structures on the B-mode image.

Related to these two artefacts, care needs to be taken to ensure that the image is not being affected by the stiffness of structures immediately outside the imaging plane, e.g. boundaries, cysts, low/high strain structures.

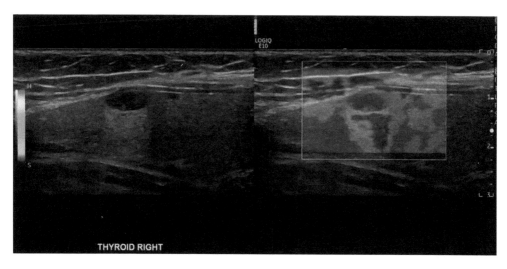

FIGURE 14.12   Example of stress concentration artefact. Elastogram of a hypoechoic thyroid nodule shows the tissue distal to the stiff (red) nodule as softer (blue) than the surrounding tissue. *Source:* Courtesy of Stephen Klarich.

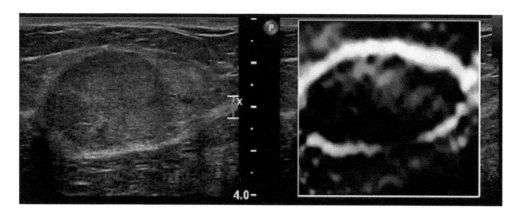

FIGURE 14.13   A greyscale elastogram of a superficial benign breast lesion slipping within surrounding fatty tissue as the probe is palpated giving rise to an artefactual bright ring of apparently soft tissue. *Source:* Barr [5]/Korean Society of Ultrasound in Medicine.

Appearance of Cysts

A watery fluid is incomprehensible (try compressing a syringe full of water). A fluid filled cyst may distort when compressed depending on the stiffness of the tissue encapsulating it. However, fluid-filled cysts do not return echoes and appear black on B-mode. In strain imaging the lack of echoes and speckle to track produces noise and artefacts. In particular they produce what is known as blue/green/red artefact.

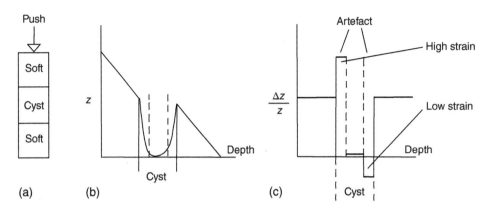

FIGURE 14.14   Illustration showing how a cyst produces 'blue/green/red' artefact in SE. (a) the tissues, (b) movement detected along the ultrasound beam, (c) the gradient of movement showing the resulting artefact.

Blue/Green/Red Artefact

Figure 14.14 shows how the gradient of strain changes at the boundaries of a non-echogenic cyst to produce blue/green/red artefact. It gets its name from the three-colour layering pattern seen in cystic lesions. Called **blue/green/red artefact** (blue = stiff), the three colours form three bands across the whole cyst as shown in Figure 14.15.

Shadowing

As with other ultrasound techniques, acoustic shadowing will cause loss of information and a noisy image in shadowed areas.

ACOUSTIC RADIATION FORCE IMPULSE IMAGING (ARFI IMAGING)

ARFI uses the third way of creating a distortion in tissue, namely to use ultrasound pulses to create a disturbance directly inside the tissue mass. When an ultrasound pulse travels through tissue, energy is absorbed from the pulse and the tissue experiences a force known as **radiation force** (see Chapter 11). For the ARFI technique a higher intensity **'push pulse'** is transmitted into the tissue, with a focused beam centred on a region of interest. This generates a disturbance in the tissue created by the radiation force from the push pulse. It produces a displacement of a few microns in the axial direction which recovers after a few milliseconds. As before, the amount of tissue movement this radiation force produces, is detected using speckle tracking or Doppler detection. By transmitting a series of these push pulses, followed by normal pulses along the same line to interrogate the tissue movement, and stepwise across

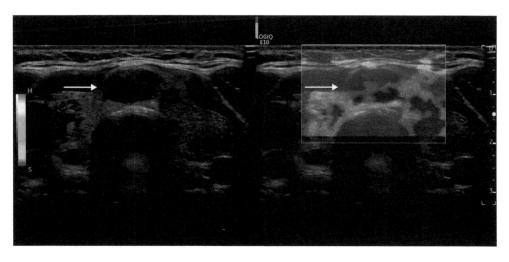

FIGURE 14.15 Illustration of 'blue/green/red' artefact in the cyst indicated by the arrow, only in this case it is red/green/blue as red = stiff in this elastogram. *Source:* Courtesy of Stephen Klarich.

the field of view, a region of interest can be scanned and the strain information be displayed using a colour mapping. A typical push pulse will be a long burst (10's of μs) of focused ultrasound.

Pros and Cons of ARFI

ARFI has a number of advantages over manual palpation.

Pros:
- Push pulses can be delivered deep into tissue.
- Delivers more uni-axial pulses which are less user dependent.
- Gives better resolution elastograms.
- Less strain-concentration artefacts.
- Cysts do not have stress cone bright ups anterior and posterior to the cyst.
- Better strain contrast and improved SNR at depth.
- Less influenced by slip movements anterior to the target of interest.

Cons:
- May be affected by the absorption and reflection of push pulses.
- Important not to push on the skin surface and pre-compress the tissue.
- The depth of imaging is limited by the maximum size of push pulse that can be safely transmitted into the body.

STRAIN RATIO

Strain imaging itself is qualitative, showing relative differences in stiffness. A semi-quantitative approach can be achieved by comparing the strain shown in a target tissue with that of a tissue whose strain is used as a normal reference value. This then forms a **strain ratio**. Most malignant tissue is stiffer than normal tissue and the strain in a suspected tumour may be compared with the strain in an adjacent area of fat within the same field of view, for example.

> **NOTES**
> - Ratios vary between scanners and values should not be compared across different manufacturers machines.
> - The reference region of interest should be positioned to the side of the target region of interest, rather than above or below it, so the pushing conditions are similar at that depth.
> - Measured ratios have greater errors than theoretical values for softer inclusions. Therefore, be very cautious when using ratios. In practice they can only reliably used to make a binary analysis, i.e. >1 harder or <1 softer.

SHEAR WAVE ELASTOGRAPHY (SWE)

Background Theory

We saw in Chapter 1 that normal ultrasound waves are longitudinal pressure waves that travel through soft tissue at the speed of sound $1540\,ms^{-1}$. The tissue will move in response to these pressure waves, but we cannot use them to directly measure elasticity because they travel too fast, traveling at the same rate as the ultrasound pulses we might use to interrogate them.

For quantitative measurement of tissue elasticity with ultrasound, the solution is to make use of **shear waves**. Consider one of those 'slinky' spring toys (Figure 14.16). If we stretch it out and move the end backwards and forwards in the direction of the line of the spring, we see a **longitudinal wave** travelling along the spring, exactly like a normal sound wave. We can induce another sort of wave in the spring by moving the end sideways relative to the axis of the spring, similar to shaking a rope from side to side. This is called a **transverse wave.** When such a transverse wave occurs within a solid it is called a **shear wave**.

> **NOTE**
> Here, transverse or shear waves are a different phenomenon from transverse oscillations used for vector flow mapping (Chapter 12). Those used for vector flow mapping are longitudinal compression waves travelling out along the x-axis.

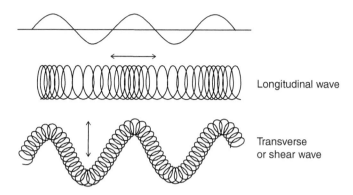

FIGURE 14.16 Longitudinal wave and transverse wave seen by oscillating a 'slinky'. In both cases the wave travels along the slinky from one end to the other.

In a solid material, such transverse oscillations of molecules push and pull on those molecules next to them and the shear wave will travel out through the material from the source of the wave (Figure 14.17). The particle displacement is perpendicular to the direction of propagation and the speed of these shear waves is about 1000 times slower than the 1540 ms^{-1} speed of sound of longitudinal waves, i.e. 1–10 ms^{-1}.

To picture what goes on in tissue and how we might detect shear waves think of a stone dropped into a still pond. The waves on the surface of water are like shear waves. Their oscillation is perpendicular to their direction of travel. The stone is the source of the waves and the waves travel out from the source with the surface going up and down. A cork sitting on the surface some distance away from the source will start to bob up and down with the expanding wave as the wave reaches it and passes it. The motion of the cork is in the same vertical direction as the stone travelled that started the waves. So, for our ultrasound case, the molecules will oscillate in the z-direction but the wave is travelling out in the x-direction. We can therefore use normal ultrasound pulses to detect this motion in the z-direction using speckle tracking, and find the time it takes for the shear wave to reach a given point away from the source and so find the shear wave velocity in the x-direction (Figure 14.18).

Within the soft tissues found in the body, shear waves do occur, but are very much weaker than the normal longitudinal waves as soft tissue cannot support transverse waves very well. However, shear waves are more sensitive to tissue stiffness than longitudinal waves and they can provide the 'probe' to investigate tissue stiffness. In particular the **speed of sound of the shear wave** can be used to differentiate the stiffness of soft tissues.

Recall from Chapter 1, the speed of sound of the normal longitudinal wave is given by

$$c = \sqrt{\frac{K}{\rho}}$$

where K is the **bulk modulus** and ρ the density of the material.

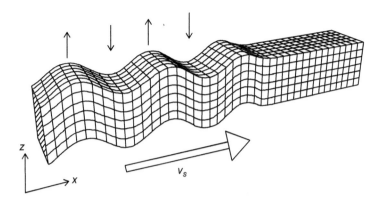

FIGURE 14.17   Diagram of a shear wave in a solid showing the transverse motion of molecules whilst the wave propagates in the direction show by the arrow.

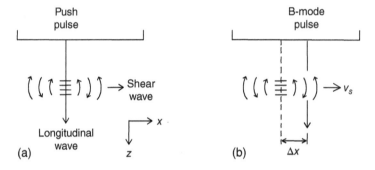

FIGURE 14.18   Diagram showing (a) the shear waves generated by longitudinal push pulse oscillating in the z direction whilst propagating outward in the x direction and (b) a B-mode pulse to interrogate the shear wave at time Δt for distance Δx to find shear wave velocity $v_s = \Delta x / \Delta t$.

The bulk modulus is the ratio of the change in change in volume $\dfrac{\Delta V}{V}$ to a change in applied pressure ΔP.

$$K = V \frac{\Delta P}{\Delta V} \left(\text{see Figure 1.20} \right)$$

In a similar manner, the speed of sound for shear waves c_s is given by

$$c_s = \sqrt{\frac{G}{\Delta}}$$

where G is the **shear modulus**, the ratio of the applied force to the resulting distortion (Figure 14.19).

$$G = \frac{\text{transverse force}}{\text{resulting distortion}}$$

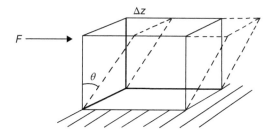

FIGURE 14.19   Definition of shear modulus showing distortion Δz caused by an applied force F.

Shear wave speeds of sound c_s in soft tissue range from 1–10 ms^{-1} and G has a large range.

Young's modulus E and shear modulus G are related

$$E = DG = D\rho c_s^2$$

where D is a constant of proportionality.

For the soft abdominal tissues $D = 3$, i.e. for these tissues

$$E = 3G = 3\rho c_s^2$$

So, if we measure the shear wave speed of sound c_s, we can estimate E and G in kPa (or Nm^{-2}) and obtain quantitative estimates of tissue stiffness.

For SWE the applied force used is acoustic radiation force 'push pulses', as used in ARFI SE. However, instead of performing speckle tracking of movement along the same beam axis that the push pulse was sent out, speckle tracking is performed along the beam axis (Δz) of B-mode beams sent out in pulses moving laterally away from the push pulse axis, as shown in Figure 14.20. As tissue is disturbed by the push pulse, shear waves are generated, disturbing the tissue laterally, away, from the axis of the

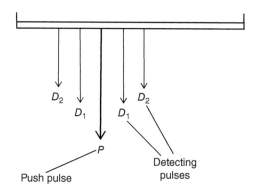

FIGURE 14.20   Illustrating the pulsing regime used for point SWE with a push pulse followed by detecting pulses to monitor the shear wave travelling away from the push pulse.

push pulse beam, with the disturbance moving out at the shear wave speed of sound. The wavelength of the shear waves is about 1 cm and depends on the width of the tissue directly excited by the push pulse [7]. The resolution of the elastogram depends on the resolution of the interrogating B-mode beams and is about 1 mm. The speed at which the shear waves move can be tracked across a band 5–10 mm wide using the B-mode pulses to detect the transverse oscillation of tissue (Δz), as the shear wave passes by.

POINT SWE (PSWE)

This technique can give a shear wave speed of sound measurement for a single uniform region of interest with a single push pulse beam and detection of resulting shear wave motion, hence **point SWE**. Each measurement is static and not updated. It may be used to give a quantitative single value measure of liver, or other homogenous organ, stiffness (Figure 14.21).

As the actual shear wave speed of sound is measured, the information may be given as the shear wave speed of sound or, using the equation

$$E = 3\rho c_s^2 \text{ for soft tissue,}$$

Young's modulus can be calculated and the information given as the stiffness (E) in kPa.

An extension of this technique is to repeat the push-pulse – detection-pulse sequence for a range of depths and across a region of interest (ROI) to give a **static shear wave elasticity image** as a colour elastogram box on the B-mode image (Figures 14.22 and 14.23). The image is static in that it is a single image that is not updated. Using this technique, the speed of shear waves can be measured to a depth of ~8 cm.

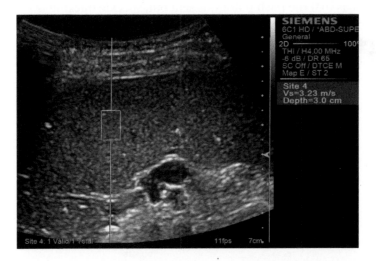

FIGURE 14.21   Point SWE in a patient with splenomegaly. The sample volume position is indicated by the cursor, the average value of shear wave speed in the sample is shown on the right. *Source:* Pereira et al. [8]/with permission of Elsevier.

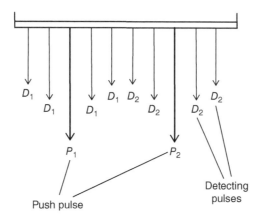

FIGURE 14.22   Typical pulsing regime across an array used to form a SWE image.

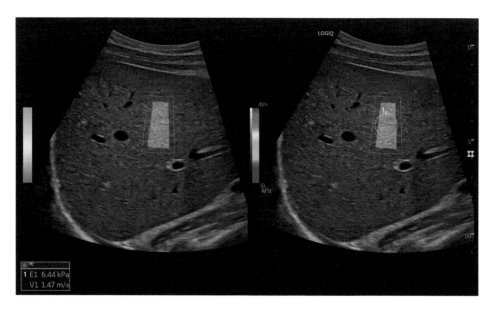

FIGURE 14.23   Shear wave elastogram of liver showing the elastogram on the right and a quality indicator on the left. Average value of stiffness within the circle shown bottom left.

A number of push pulses are transmitted, focusing at different depths within the elastogram box and across its width. As these deliver a significant amount of energy to the tissue there needs to be a rest period between each refresh sequence. This is in order to keep the tissue exposure within safe limits and to prevent the transducer from overheating (see Chapter 11). The frame rate is therefore determined by the acquisition time and the cooling time. Using multiple simultaneous transmit pulses across the ROI and interpolation of acquired values, the elastogram can be refreshed at a faster rate. Typically, the acquisition time may be 100 ms and the cooling time 2–3 seconds, meaning the cooling time is the dominant factor in determining frame rate.

As with SE, a quality indication may be displayed to allow the user to gauge the reliability of the elastogram.

NOTES

- For other tissues, e.g. kidney and muscle D does not equal three and in these cases the shear modulus G can be estimated and used instead of Young's modulus.
- The shear modulus G also has units kPa.
- For all SWE techniques the results are less accurate for depths below 6cm due to attenuation of the push pulse generating the shear waves.

SUPERSONIC SHEAR IMAGING (SSI)

This is the most advanced form of shear wave imaging and gives a real-time 2-D or 3-D map of tissue stiffness refreshed several times a second.

As for conventional SWE, SSI uses a series of ARFI push pulses to generate shear waves. A set of push pulses are sent out along a single transmit beam focused at increasing depths along that line (Figure 14.24). The rate of transmission of these push pulses is such that the foci, generating shear waves, move along the transmit line faster than the speed of the shear waves themselves. Hence the name – **supersonic shear imaging**. This creates a column of shear waves moving out from along the whole length of the push pulse beam in the shape of a cone called a **Mach cone**, in the manner of a wake seen behind a moving boat or a sonic boom moving out behind a supersonic plane. The shear waves produced are then interrogated by speckle tracking pulses using the ultrafast ultrasound technique, for example, plane wave imaging with post transmit-receive beam forming (Chapter 13).

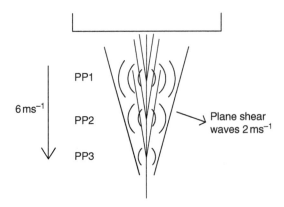

FIGURE 14.24   Using a series of push pulses PP1–3 generated at supersonic speed to produce a Mach cone of shear waves.

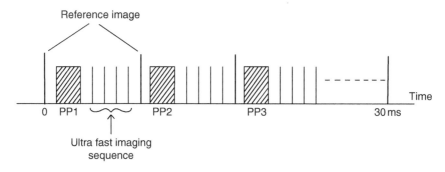

FIGURE 14.25 A typical push pulse (PP) sequence used for SSI, showing the B-mode reference images, push pulse sequences and imaging sequences detecting the shear wave velocities.

The following pulsing sequence is typical (Figure 14.25). A reference B-mode image is created to give the static tissue state against which the shear wave movement is compared. A series of push-pulses are sent out to create a Mach cone of shear waves moving out from the beam axis. The shear waves are then tracked using speckle tracking from beams adjacent to the push-pulse beam axis using ultrafast plane-wave imaging. This sequence is repeated for a series of push-pulse lines across the image plane to cover the region of interest. The speed of sound data for shear waves is used to calculate stiffness and elastograms are produced at the rate of 3-4 Hz. Frame averaging is used to reduce noise (Figure 14.26).

SHEAR WAVE COMPOUNDING

The ratio of the speed of travel to the speed of sound is called the **Mach Number**, for example, a plane travelling at twice the speed of sound is travelling at Mach 2. The angle at which the shear wavefront moves away from the transmit line depends on the Mach number of the transmit push pulses along the transmit line. By generating transmit lines with different Mach numbers, shear waves are generated at different angles and the elastograms from these transmissions can be averaged to give **shear wave compounding** (Figure 14.27). This is similar to B-mode spatial compounding (see Figure 7.6).

NOTES

- Shear wave elastography is good for assessing **diffuse disease** as it gives quantitative values from areas in which the change in stiffness from one point to another is small. In other words, there are no normal adjacent regions for comparison.
- As it is quantitative, it allows changes over time to be evaluated.
- SSI is insensitive to patient motion due to the fast acquisition time

- There is a limitation on the depth of field of view (typically 6 to 8 cm) as it is difficult to safely generate shear waves in deep tissue and the signal gets noisy. The push pulse intensity would have to be too high to go deeper.
- Caution should be applied when comparing results across different manufacturers scanners. [10, 11]
- For shear wave elastography the probe should be held completely still with little pressure on the skin (see Figure 14.10).
- Fluids do not support shear waves. Cysts and other focal lesions tend to produce artefacts on shear wave images, as described above. For this reason, vessels should be avoided within the image.

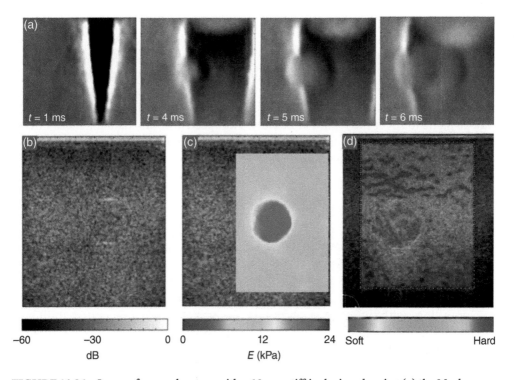

FIGURE 14.26   Images from a phantom with a 10 mm stiff inclusion showing (a) the Mach cone arising from a set of supersonic push pulses at increasing time intervals after generation. The stiff inclusion is seen to distort the shear wavefront. (b) a normal B-mode image of the phantom where the inclusion is hardly visible. (c) The shear wave elastogram of the inclusion and (d) a strain elastogram of the same phantom using manual palpation of the probe. *Source:* Fink and Tanter [9]/with permission of AIP Publishing LLC.

The estimate of the shear wave velocity v_s depends on reliable acquisition (absent patient/probe movement) and reliable data processing (low SNR). Based on detecting movement and noise from the ultrasound data, a quality measure may be estimated for each sample position to create a 2D quality map that can guide the user in assessing the quality/reliability of the elastogram shown (Figure 14.28).

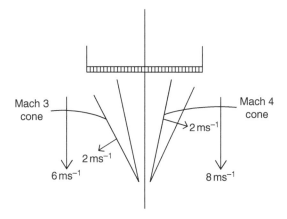

FIGURE 14.27 Diagram showing the change in Mach cone angle as the Mach number increases from 3 to 4.

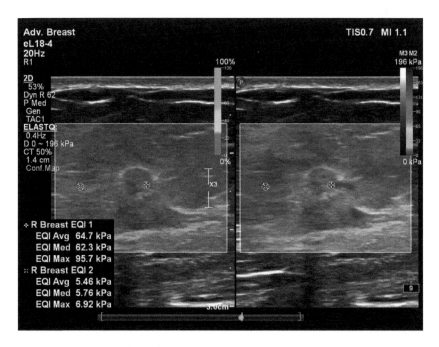

FIGURE 14.28 SWE Elastogram of the breast on right with elastogram quality indicated on the left. Stiffness, measured as Youngs Modulus, at the two measurement sites shown has been derived from a set of repeated measurements. *Source:* Illustration by courtesy of Koninklijke Philips N.V.

SWE ARTEFACTS

Bang Artefact

The appearance of a stiff area in the near field caused by pushing down on the skin with the transducer creating pre-compression of the tissue (Figure 14.29). This may be avoided by use of coupling gel and less pressure on the transducer probe.

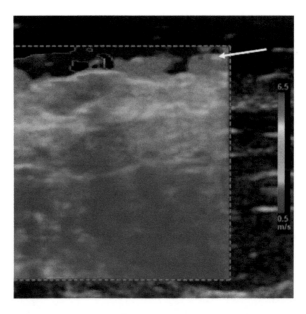

FIGURE 14.29   Example of the bang artefact in a SWE image of a breast where too much pressure is applied to the skin giving a falsely high value at the top of the elastogram.
Source: Barr [12]/Korean Society of Ultrasound in Medicine.

REFERENCES

1. Bamber, J., Cosgrove, D., Dietrich, C.F. et al. (2013). EFSUMB Guidelines and recommendations on the clinical use of ultrasound elastography. Part 1: basic principles and technology. *Ultraschall in der Medizin* 34: 169–184.

2. Shiina, T., Nightingale, K.R., Palmeri, M.L. et al. (2015). WFUMB guidelines and recommendations on the clinical use of ultrasound elastography: Part 1: basic principles and terminology. *Ultrasound in Medicine and Biology* 14: 1126–1147. https://doi.org/10.1016/j.ultrasmedbio.2015.03.009.

3. Dietrich, C.F., Barr, R.G., Farrokh, A. et al. (2017). Strain elastography - how to do it? *Ultrasound International Open* 3: E137–E149.

4. Barr, R.G. and Zhang, Z. (2012). Effects of precompression on elasticity imaging of the breast. *Journal of Ultrasound in Medicine* 31: 895–902.

5. Barr, R.G. (2019). Future of breast elastography. *Ultrasonography* 38: 93–105. https://doi.org/10.14366/usg.18053.

6. Dietrich, C.F., Barr, R.G., Farrokh, A. et al. (2017). Strain elastography - how to do it? *Ultrasound International Open* 3: E137–E149. https://doi.org/10.1055/s-0043-119412.

7. Nightingale, K. (2011). Acoustic radiation force impulse (ARFI) imaging: a review. *Current Medical Imaging Reviews* 7 (4): 328–339. https://doi.org/10.2174/157340511798038657.

8. Pereira, C.L.D., Santos, J.C., Arruda, R.M. et al. (2021). Evaluation of schistosomiasis mansoni morbidity by hepatic and splenic elastography. *Ultrasound in Medicine and Biology* 47: 1235–1243. https://doi.org/10.1016/j.ultrasmedbio.2021.01.022.

9. Fink, M. and Tanter, M. (2010). Multiwave imaging and super resolution. *Physics Today* 63: 28–33. https://doi.org/10.1063/1.3326986.

10. Suh, C.H., Yoon, H.M., Jung, S.C. et al. (2019). Accuracy and precision of ultrasound shear wave elasticity measurements according to target elasticity and acquisition depth: a phantom study. *PLoS One* 14: e0219621. https://doi.org/10.1371/journal.pone.0219621.

11. Yoo, J., Lee, J.M., Joo, I. et al. (2020). Assessment of liver fibrosis using 2-dimensional shear wave elastography: a prospective study of intra- and interobserver repeatability and comparison with point shear wave elastography. *Ultrasonography* 39: 52–59. https://doi.org/10.14366/usg.19013.

12. Barr, R.G. (2019). Future of breast elastography. *Ultrasonography* 38: 93–105.

Appendix 1: Knobology

'Knobology' is the term that is frequently used to describe the operation of a scanners controls in order to obtain the optimal image in an ultrasound examination. There are standard settings that would generally be applied when imaging a particular area, for example, abdomen, neck, muscular-skeletal, and when using particular modalities, for example, B-mode, Doppler, elastography. In addition to the standard settings, controls need to be adjusted throughout an examination to optimise the image and make measurements.

Modern scanners have presets which set the scanner up for a particular examination using standard settings. These alter a large number of the controls. In addition, techniques to automatically assess the image and then optimise it during the course of an exam are being introduced. Some of the more portable scanners designed for point-of-care use have very limited user access to controls. Presets and auto-adjustments often do a very good job of managing image quality. However, it is important that you as the sonographer are aware of what is being done and that you know how to manually adjust the scanner to give the best results you can, for the patient in front of you. There will always be situations where a good image is hard to obtain. Careful adjustment of the controls is then needed to optimise the image to give the diagnostic information required from the scan.

All of the controls and what they do have been described in this book. In this appendix, they are presented in the form of a checklist, much in the way an aircraft pilot will have a checklist to ensure their controls are working and in the correct position for safe flight.

PRE-SCAN

- Is the room lighting suitable for scanning and viewing images?
- Is the monitor correctly set up for scanning so that it shows the full range of greyscale?
- Is the patient couch and sonographer seat adjustable and set up for safe scanning to avoid repetitive strain injury (RSI) problems for the sonographer?

B-MODE SCANNING

- Choice of **probe** – can you justify the choice of ultrasound probe for the current exam?
  - Probe type
  - Choice of frequency for resolution vs penetration
  - Is the probe clean?

Pre-Scan

- Are the **presets** set for the current exam?
- Is the correct **Thermal Index (TI)** selected for the current exam, e.g. TIS, TIB, TIC?
- Is **transmit power** adjusted to a safe level whilst able to achieve adequate penetration?
  - Are there special factors such as scanning the eye, intraoperative or endoscopic exam of early pregnancy to be taken into account?
- Are the **TI** and **MI** values displayed on the screen showing safe operating levels?
- Are the **greyscale mapping** (transfer function) and **dynamic range** set appropriately?
- Are you using
  - **Compound scanning**
  - **Adaptive processing**
  - **Harmonic imaging**
  - Can you justify and explain their use?

During the Scan

- Adjust the **gain** and time gain compensation(**TGC**) to show a uniform image at all depths
- **Focus**
  - Set the focus to the depth of the target of interest
  - Are you using multiple transmit focuses?
- **Frame rate**
  - Is the frame rate adequate?
    (affected by multiple transmit focuses, field of view size)
- **Frame averaging** – adjust to reduce blurring or to improve image quality

Making Measurements

- Correct choice of measurement and correct placement of calipers.
- Does the measurement need to be repeated?

CONTRAST-ENHANCED ULTRASOUND (CEUS)

- Awareness of sensitive tissues and increased risk of cavitation (cf **MI**)
- Use of **harmonic imaging**

DOPPLER WAVEFORMS

Pre-Scan

- Are the **presets** set for the current exam?
- Is the correct **Thermal Index (TI)** selected for the current exam, e.g. TIS, TIB, TIC?
- Is **Doppler transmit power** adjusted to a safe level whilst able to achieve adequate penetration?
- Are the **TI** and **MI** values displayed on the screen showing safe operating levels?

During the Exam

- Is the **sample volume** length and position set correctly on the image?
  - Can you justify the choice?
- **Doppler angle**
  - Is the vessel correctly aligned?
  - Is the angle cursor correctly aligned to the vessel?
  - Is 'heel and toeing' required to achieve an adequate angle
  - Can you justify the angle used?
- Is the **wall filter** setting adequate?
- Is the image frozen whilst the Doppler waveform is acquired?
- Is the **Doppler gain** set to show full black to white on the sonogram with no over or under gain?
- Does the waveform need **inverting**?
- Is the **scale** set correctly so that the waveform fills the displayed range without aliasing?
- Is the **sweep speed** (time base) set correctly to show the relevant number of waveforms and no more?

Measurements

- Is **autotrace** or automatic caliper placement being used?
- Are the measurements being presented consistent with the observed waveform? (e.g. autotrace is not picking up noise, venous signal)
- Should manual placement of the cursors be used?
- Is a repeated measurement needed?

COLOUR DOPPLER ULTRASOUND (CDU)

Pre-Scan

- Are the **presets** set for the current exam?
- Is the correct **Thermal Index (TI)** selected for the current exam, e.g. TIS, TIB, TIC?
- Is **CDU transmit power** adjusted to a safe level whilst able to achieve adequate penetration?
- Are the **TI** and **MI** values displayed on the screen showing safe operating levels?

During the Scan

- Does the **scale** need to be inverted to show the target of interest correctly?
- Adjust the **colour gain** so that colour is shown without colour bleeding or dropout
- Is the **wall filter** setting adequate?
- Is the **B-mode priority** adjusted correctly?
- **Frame Rate**
  - Adjust the size of the colour box to give an adequate frame rate
  - Adjust the colour frame averaging to avoid blurring whilst improving image quality
- Adjust the **scale** to avoid aliasing (or to use it diagnostically)
- Is 'heel and toeing required to achieve an adequate angle or avoided to increase sensitivity?

ELASTOGRAPHY

Pre-Scan

- Are the **presets** set for the current exam?
- Is the correct **Thermal Index (TI)** selected for the current exam, e.g. TIS, TIB, TIC?
- Are the **TI** and **MI** values displayed on the screen showing safe operating levels?

During the Scan

Strain Elastography

- Is the **elastogram ultrasound frequency** set for adequate penetration vs resolution?
- Set **colour map** for red/blue vs hard/soft
- **Frame Rate**
  - Adjust **frame rate control**
  - Lower **elastogram ultrasound frequency** lowers frame rate
  - Reduce **line density** increases frame rate
- Adjust **tracking window** for sensitivity (improved SNR) vs resolution
- Adjust **frame reject** to balance noisy frames vs elastogram quality (affects flashing of elastogram)
- Adjust **noise reject** to balance noise vs pixel dropout
- Adjust **frame averaging** to improve image quality and reduce flashing
- Adjust **dynamic range** to determine level of significant colour change shown
- Adjust **colour blend** for transparency of elastogram over B-mode display

Shear Wave Elastography

- Be aware of **TI** and **MI** values exceeding safe limits.

Appendix 2: Handling Equations and Decibels

EQUATIONS

Equations may seem daunting and using them to calculate values not very familiar. Their use within this book has been kept to a minimum, but it is not really possible to grasp the essentials of ultrasound technology without including some equations and showing some numerical results. This appendix has been included as a reminder of how to manipulate equations to get the results we want to know.

It may be helpful to think of an equation as a special kind of diagram, composed of numbers, or symbols representing numbers, that shows the **relationship** of one item to another. The symbols often represent both a number and a property in the real world.

For example velocity $v_0 = 5v_1$

The property is the velocity of something and this equation tells us that whatever value v_0 has, v_1 will have 5 times that value.

Each side of an equation has a top and a bottom, but where the bottom equals 1, we do not usually write it.

$$\text{Examples:} \quad \frac{v}{1} = v \text{ or } \frac{AB}{1} = \frac{D}{E} + \frac{F}{1} \text{ normally written } AB = \frac{D}{E} + F$$

An equation is like a 'seesaw' type of weighing scales where one side must **always** balance or equal the other. In order to make sure you have the right result you can write out every stage in manipulating the equation, but there are a number of short cuts (rules) you can use as shown in the examples below.

Rule 0 You can change things around on an equation but whatever you do to one side you must also do to the other, so as to keep it in balance.

Adding or Subtracting $A = B$ subtract B from $A \Longrightarrow A - B = B - B = 0$

$$\left(-B \text{ on both sides}\right)$$

i.e. $A - B = 0$ and note how B has changed sides. This gives:

Rule 1   If an item moves from one side to the other it changes sign.

Dividing $A = B \Longrightarrow \dfrac{A}{B} = \dfrac{B}{B} = 1$ (divide both sides by B) This gives:

Rule 2   If the same item appears on the top and the bottom of one side, we can cancel them out and replace them with 1.

and

Rule 3   We can move an item from the top of one side to the bottom of the other side (and vice versa) and the equation stays in balance.

Example $\dfrac{A}{B} = \dfrac{C}{D} \Longrightarrow \dfrac{AD}{B} = \dfrac{CD}{D} = C$ (D moves from the bottom to the top)

Multiplying   Top and bottom of one side by the same amount.

Example $A = B \Longrightarrow A = \dfrac{BB}{B} = \dfrac{B^2}{B}$

Here we could cancel the B's on the right to get back to where we started, but it is sometimes useful when rearranging equations to do the same thing to the top and bottom of one side and leave it there before doing something else. This gives:

Rule 4   You can do the same thing to the top and the bottom of one side and the equation will stay in balance.

Multiplying and dividing   when there are several terms

Example $A + B = C \Longrightarrow A.\ D + B.\ D = C.\ D = D(A + B)$

D was a new item but we multiplied it to all terms on both sides so the equation stays in balance. This gives:

Rule 5   Where there are several terms (separated by + or −) on one side, then you must do the same to each of the terms.

and

Rule 6   Where the same item is in each term then the terms can be collected together in parentheses with the common item outside. This also means that to expand such parentheses, each term inside the parentheses is multiplied by each item outside the parentheses.

$$\text{Example } A + B = C \Rightarrow \frac{A}{C} + \frac{B}{C} = \frac{C}{C} = 1 = \frac{A + B}{C}$$

A and *B* terms are each divided by the same amount C. This gives:

Rule 7   If all terms on one side are divided by the same item, we can collect them together over that item (but *only if* they have the same item on the bottom).

$$\text{Example } \frac{A}{C} + \frac{B}{CD} \neq \frac{A + B}{C} + \frac{B}{D}$$

i.e. WRONG because on the bottom of the left side not all terms are the same.

$$\text{However } \frac{A}{C} + \frac{B}{CD} = \frac{AD}{CD} + \frac{B}{CD} = \frac{AD + B}{CD}$$

by multiplying top and bottom of the first term by *D* all terms then have *CD* on the bottom.

Rule 8   Anything in parentheses can be treated as a single item.

$$\text{Examples } E = D\left(A + C\right) \Rightarrow \frac{E}{\left(A + C\right)} = D$$

parentheses moved as per Rule 3

$$E = D\left(\frac{AB}{C}\right) \Rightarrow \frac{E}{\left(\frac{AB}{C}\right)} = D$$

Here $\left(\dfrac{AB}{C}\right)$ has been treated as a single item and moved to the bottom of the other side as per Rule 3. Then:

Rule 9 $\dfrac{1}{\text{fraction}}$ is the same as inverting the fraction on the bottom and putting it in the normal position

Example $\dfrac{1}{\left(\dfrac{A}{B}\right)} = \dfrac{B}{\left(\dfrac{AB}{B}\right)} = \dfrac{B}{A}$

here we multiply top and bottom by B and the B's on the bottom then cancel.

Tip If you want to check that you have done the right thing in manipulating an equation, then put in some numbers and see if it gives the right answer.

Example $A = 2, B = 3, C = 4, D = 5$ $\quad \dfrac{A}{C} + \dfrac{B}{CD} = \dfrac{2}{4} + \dfrac{3}{20} = 0.65$

rearrange the equation $\qquad\qquad \dfrac{AD + B}{CD} = \dfrac{10 + 3}{20} = 0.65$

Doppler Equation as an Example

The equation may be remembered as shown on the left. It gives the Doppler frequency from a target velocity v, speed of sound c, the transmitted frequency f_T tissue and the Doppler angle θ_D.

$$f_D = 2\dfrac{v}{c} f_T \cos\theta_D \quad V = \dfrac{f_D}{f_T} \dfrac{c}{2\cos\theta_D}$$

The equation shows the relationship between these quantities and we see that for a given transmit frequency the Doppler frequency depends on the ratio of the target velocity to the speed of sound in tissue. It also changes as the cosine of the Doppler angle. We know that $\cos 90° = 0$, so a Doppler angle of 90° gives a Doppler frequency of 0 regardless of the target velocity.

The thing we want to know is the target velocity, say, blood velocity, in the body. On the right, the Doppler equation has been rearranged using Rule 3 to show the relationships in terms of the target velocity. Once we know the Doppler frequency for a given angle, the blood velocity producing that frequency can be calculated and we see that the value depends on the ratio of the Doppler frequency to the transmitted frequency.

Tips for Calculating Answers to Problems

1. Make sure all of the units are consistent

e.g. kilograms (kg), metres (m), seconds (s)

Do not mix, for example, millimetres with metres in the calculation or you will end up with errors in factors of 10.

2. Make sure you have calculated all the factors of 10 correctly

Factors of 10 can often be the source of errors in the answer. Multiplying a number by a power of 10 is the same as moving the decimal point that number of places to the right, and dividing by a power of 10 is the same as moving the decimal place to the left.

$$\text{Examples } 15.72 \times 100 = 1572.0$$

this can also be written as 1.572×10^3 that is, as a unitary number multiplied by a power of 10.

$$5.78/1000 = 0.00587 = 5.78 \times 10^3$$
$$0.0733 = 73.3 \times 10^3$$

Note:
- Anything to the power of 1 is itself, e.g. $4^1 = 4$
- Anything to the power of 0 is 1, e.g. $10^0 = 1$, $6^0 = 1$

To multiply powers of 10 add the indices

$$\text{e.g. } 10^6 \times 10^4 = 10^{10}$$

To divide powers of 10 subtract the indices

$$\text{e.g. } \frac{10^2}{10^5} = 10^{-3} = \frac{1}{10^3}$$

Note:
- it is always 'top minus bottom'
- a power can move from top to bottom with a change in sign

$$\text{e.g. } \frac{1}{100} = 10^{-2} = 0.01 = 1 \times 10^{-2} \text{ but we usually omit the '1'}$$

TABLE A2.1   Units used in expressing the values of variables.

Power	Name	Symbol
10^6	mega	M
10^3	kilo	k
10^1	10	
10^0	1	
10^{-2}	centi	c
10^{-3}	milli	m
10^{-6}	micro	μ
10^{-9}	nano	n

To make sure you have calculated the powers of 10 correctly, it is a good idea to write all the numbers as unitary numbers with powers of 10 and then calculate the result from the numbers and then add up the powers of 10.

$$\text{e.g. } \frac{500 \times 0.0072}{60 \times 10^4 \times 0.01} = \frac{5 \times 10^2 \times 7.2 \times 10^{-3}}{6 \times 10^5 \times 10^{-2}} = 6 \times 10^{-4}$$

if this was distance in metres then the answer in millimetres $= 0.6\,\text{mm}$, i.e. divided by 1000

3. Ask yourself the question 'is this a sensible answer?'

   e.g. blood velocity $= 500\,\text{ms}^{-1}$ probably wrong by a factor or 100 or 1000!

4. To get the units (e.g. kg, m, s) right, do the calculation of the numbers. Then write the equation with just the units and rearrange the equation to see in what units the numerical answer should be given.

   Example: In Chapter 1 we saw that the speed of sound in a medium was given by the square root of the bulk modulus κ divided by the density ρ

   Where $K = -V\dfrac{\Delta P}{\Delta V}$ has units $\text{kgm}^{-1}\text{s}^{-2}$ and density ρ has units kgm^{-3}

   Then, looking at the units $c = \sqrt{\dfrac{K}{\rho}} = \sqrt{\dfrac{\text{kg m}^{-1}\text{s}^{-2}}{\text{kgm}^{-3}}} = \sqrt{\text{m}^2\text{s}^{-2}} = \text{ms}^{-1}$

   Which is the correct units for a speed of sound. Note how the units in an equation behave in the same way as any symbols or numbers.

Final Example Combining All These Rules and Tips

Question

If the speed of sound in a plastic is $4.7\,\text{kms}^{-1}$ and a reflector is $3\,\text{cm}$ away from the transducer, what is the pulse-echo time from the reflector?

Answer

1. Find the right equation (draw a diagram if it helps, Figure A2.1 shows the pulse-echo principle with the variables to be determined indicated)

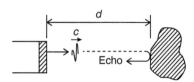

FIGURE A2.1   The pulse-echo principle.

$$\text{velocity} = \frac{\text{distance}}{\text{time}} \text{ (think of speed = miles per hour) so the equation we need is}$$

$$c = \frac{2d}{t} \text{ where the 2 is for go and return distance } 2d.$$

2. Rearrange the equation to get it in the right form to give the answer

$$t = \frac{2d}{c}$$

3. Check all the units are consistent
   $$c = 4.7 \text{ kms}^{-1} = 4.7 \times 10^3 \text{ ms}^{-1}$$
   $$d = 3 \text{ cm} = 3 \times 10^{-2} \text{ m}$$
   everything is now in metres and seconds
4. Do the calculation

$$t = \frac{2 \times 3 \times 10^{-2}}{4.7 \times 10^3} = 1.28 \times 10^{-5} \text{ s} = 12.8 \,\mu\text{s}$$

5. Is this a sensible answer?
   Recall that for soft tissue the average speed of sound is 1540 ms^{-1} and go and return time is $13 \,\mu\text{s}$ per centimetre. Here the speed of sound is about 3×1540 and the distance is 3 cm so $12.8 \,\mu\text{s}$ is the right answer.

DECIBEL NOTATION

We often need to measure the **ratio** of change in energy, power or intensity

e.g. - proportion of energy reflected at an interface
       - amount absorbed
       - gain of an amplifier e.g. 1 : 1000

There is a compact way of writing and calculating such ratios that avoids handling large numbers. To do this we take the logarithm of the ratio. This then gives the change as **Decibels (dB)**

defined as: number of dB $= 10. \log_{10}$ (ratio)

e.g.: $10. \log_{10} (2/1) = +3\text{dB}$

a ratio of a doubling, i.e. 2 : 1 has a dB value very close to 3dB

conversely: ratio $= \text{antilog}_{10}$ (no. of dB/10) which for +3dB gives a ratio of 2

Advantages

- whereas ratios multiply or divide, decibels add or subtract

$$\text{e.g. multiply } \frac{1}{2} \times \frac{1}{2} = \frac{1}{4} \text{ but add } 3\text{dB} + 3\text{dB} = 6\text{dB}$$

- it is easy to express very large ratios as small numbers.

$$\text{e.g. } 10\,000 : 1 = 40\text{dB}$$

So:	ratio 2	corresponds to	$+3\text{dB}$
	ratio 10	corresponds to	$+10\text{dB}$
	ratio 0.5	corresponds to	-3dB
	ratio 0.1	corresponds to	-10dB

NOTE

Where the ratio is fractional, e.g. 0.5, the decibel value is negative.

Now we can easily convert a ratio using multiples of 2 and 10

$$\text{e.g. } 1 : 4000 = 2 \times 2 \times 10 \times 10 \times 10$$

$$= 3 + 3 + 10 + 10 + 10 = 36\text{dB}$$

Conversely $-60\text{dB} = -30 - 30 = 10^{-3} \times 10^{-3} = 10^{-6}$

Note the similarity to adding indices for multiples of 10 or using a logarithmic scale on a graph to show a very large range of values.

or by direct calculation

$$\text{e.g. } 15\% = 0.15 = 10.\log_{10}\left(0.15\right) = -8.24\text{dB}$$

IMPORTANT NOTE

Sometimes dB are used to express the ratio of two pressures or voltages. These are values of the **amplitude** of the signal. Since power or intensity is $\propto$ (amplitude)^2 i.e. p^2 or V^2, then **for expressing amplitudes as dB we must use**:

$$dB = 20 \log_{10}\left(\text{amplitude}\right) \text{ so a doubling of } amplitude = +6dB$$

$$\text{a ratio of } 10 \text{ in } amplitude = +20dB, \text{etc.}$$

Appendix 3: The Unfocused Transducer Beam Shape

If we look at a field point in the ultrasound beam, we can measure the geometrical distance to various points across the transducer aperture to determine whether there will be constructive or destructive interference at that field point. The easiest way to think about it is to do the reverse, to consider an echo arising from the field point and to look at what the transducer 'sees' of the phase from that echo. In this Appendix, we consider a plane unfocused a transducer with continuous wave transmission at a single frequency and consider the receive ultrasound beam, i.e. the one-way beam. The transmit beam will have the same shape and the go-and-return beam shape will be the combination of the two beams. Several field points that we will consider are numbered on the Figure A3.1.

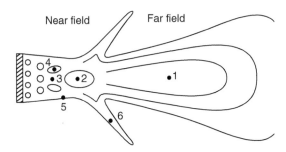

FIGURE A3.1   Intensity contour map of a continuous wave ultrasound beam from a plane aperture. Index points indicated.

A POINT IN THE FAR FIELD (FIGURE A3.2)

In the far field, field point (1) – beyond the last axial maximum – the difference in path length between the centre of the aperture and the edge of the aperture is less than a wavelength λ and the wavefronts approximate plane waves. Another way of

Ultrasound Technology for Clinical Practitioners, First Edition. Crispian Oates.
© 2023 John Wiley & Sons Ltd. Published 2023 by John Wiley & Sons Ltd.

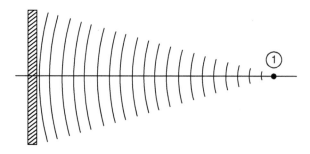

FIGURE A3.2   Wavefronts arriving at the transducer from a point in the far field.

putting it is to say that the whole of the aperture 'sees' the same phase of the wave-front at the same time from that field point.

LAST AXIAL MAXIMUM (FIGURE A3.3)

At field point (2), the edge of the aperture has a path length of just $\lambda/2$ different to the centre of the aperture. This means each half of the aperture 'sees' a positive phase at time t_0, so every point across the aperture contributes positively to the net amplitude seen at (2). Half a cycle later, each half of the aperture will see a negative going phase. As the phase of the echo across the whole aperture adds up (constructive interference), this gives a strong intensity at field point (2), i.e. the last axial maximum.

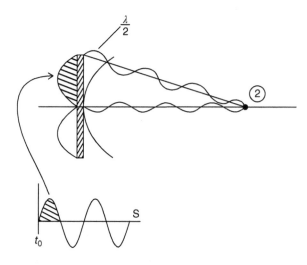

FIGURE A3.3   Wavefronts arriving at the transducer from the last axial maximum.

POINTS IN THE NEAR FIELD (FIGURE A3.4)

At field point (3), on the central axis of the beam, the path length between the centre and edge of the aperture has a phase difference of one whole wavelength. So each half of the aperture 'sees' equal amounts of positive and negative phase. These cancel out (destructive interference) and the intensity at (3) is zero.

At field point (4) each half of the half-aperture 'sees' a positive phase so the intensity at (4) is relatively high. This gives one of the peaks seen in the near field.

Going through this process for all points in the beam shows how the complex pattern of intensity in the near field arises.

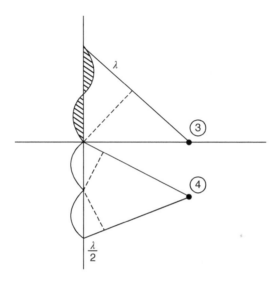

FIGURE A3.4   Wavefronts arriving at the transducer from a point in the near field (3) and field point (4).

EDGE OF THE BEAM (FIGURE A3.5)

At field point (5) each half of the aperture 'sees' the opposite sign of phase, so across the full width there is destructive interference giving a result of zero intensity, i.e. the edge of beam.

NOTE

The edge of beam (5) is not on the line of the geometrical parallel beam seen in the first simplified beam shape diagram (Figure 3.2), but is towards the centre-line of the beam. In other words, diffraction gives a natural focusing to the beam at the last axial maximum and the beam there is ~25% narrower than the parallel beam approximation.

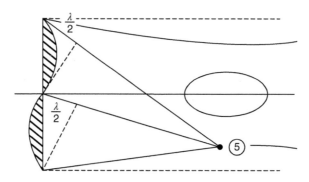

FIGURE A3.5   Wavefronts arriving at the transducer from the edge of the beam.

SIDE LOBE (FIGURE A3.6)

At field point (6), each half of the aperture again 'sees' a net positive phase so there is an increase in intensity, giving the first side lobe. However, because only 1/3 of the aperture is giving a positive contribution (2/3 cancels out), the intensity is much less than that of the main lobe and the intensity of the first side lobe is -18dB below main lobe.

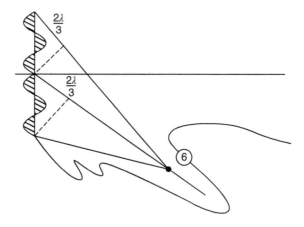

FIGURE A3.6   Wavefronts arriving at the transducer from the first sidelobe.

SUMMARY

What we see described in the intensity contour map is the diffraction pattern of a plane transducer with plane waves of a single frequency. A real beam plot of a such a transducer is seen in Figure 3.4.

Index

Ultrasound Technology for Clinical Practitioners, First Edition. Crispian Oates.
© 2023 John Wiley & Sons Ltd. Published 2023 by John Wiley & Sons Ltd.